Sequential Analysis and Optimal Design

CBMS-NSF REGIONAL CONFERENCE SERIES
IN APPLIED MATHEMATICS

A series of lectures on topics of current research interest in applied mathematics under the direction of the Conference Board of the Mathematical Sciences, supported by the National Science Foundation and published by SIAM.

HERMAN CHERNOFF
Harvard University
Cambridge, Massachusetts

Sequential Analysis and Optimal Design

SOCIETY FOR INDUSTRIAL AND APPLIED MATHEMATICS
PHILADELPHIA

Contents

Preface

This paper was prepared for a series of lectures at a Regional Conference in the Mathematical Sciences at Las Cruces, New Mexico, during December, 1972. This conference was organized by Professor G. S. Rogers and sponsored by the Conference Board in the Mathematical Sciences and the New Mexico State University with support from the National Science Foundation.

This series of lectures in design of experiments and sequential analysis is to some extent directed toward an audience of mathematicians with no knowledge of statistics. By exploring these interrelated fields, principally in the direction of my own research over a long period of time, I hope to give the audience a feeling for the nature of theoretical statistics and some idea of how it relates to the philosophy and practice of statistics. The world of applied statistics will be treated most superficially here, though it is my conviction that the attempt to answer applied problems honestly has been the major source of inspiration in theory.

At first the amount of probability theory required will be negligible, but before long there will be some slight use of the central limit theorem, the moment generating function, conditional expectation, martingales and Wiener processes. Thus an initial pretense at having these lectures be self-contained from the point of view of probability will evaporate sooner or later.

The function of statistics is to analyze results of empirical experiments in order to make appropriate decisions. The efficient selection of experiments which yield informative data is the subject of the theory of design of experiments. Sequential analysis deals with the analysis of data which are obtained in sequence. After each result arrives, the investigator must decide whether to pay the cost for additional data or whether to stop experimentation and make an appropriate terminal decision. In the act of deciding whether or not to gather more data, the statistician is making a choice of experimental design. In this sense sequential analysis relates closely to experimental design and is not really a separate field.

In the latest printing of this monograph, I have taken the opportunity to correct a number of errors that appeared in the original, hastily prepared manuscript and survived the first revised reprinting. I would like to thank Yi-Ching Yao for his assistance in finding errors in the manuscript and among the proposed corrections.

Perhaps it is the pride of authorship, but after more than 20 years, the text is still fresh and interesting to me despite the lack of reference to all of the many developments in design and optimal stopping since 1972. In particular, I should emphasize the lack of reference to the Gittins Index in bandit problems.

HERMAN CHERNOFF

Sequential Analysis and Optimal Design

Herman Chernoff

1. Preliminaries on probability. A *probability space* is a measure space $(Z, \mathscr{B}, P)$, where $\mathscr{B}$ is a sigma field (Borel Field) of subsets of Z and $P(E)$ is a countably additive nonnegative measure defined for $E \in \mathscr{B}$ for which $P(Z) = 1$. In the *frequency interpretation* of probability Z is a set of possible outcomes of an experiment and $P(E)$ is regarded as the long run proportion of times that the outcome of the experiment will fall in E when the experiment is repeated many times under similar circumstances. A *subjective probability interpretation* may be based on the odds an individual will accept in a bet whose payoff depends on whether the outcome falls in E. The set $E \in \mathscr{B}$ is called an *event*, and the event is said to be successful if the outcome falls in E.

A real-valued *random variable* X is a real-valued measurable function $X(z)$ defined on $(Z, \mathscr{B}, P)$. The probability distribution of X is the measure induced on the real line by this transformation. It is characterized by the *cumulative distribution function* (c.d.f.)

$$(1.1) \qquad F(x) = P(X \leqq x),$$

where $(X \leqq x)$ is a common abbreviation for $\{z : X(z) \leqq x\}$. The class of c.d.f.'s is the class of monotone nondecreasing right continuous functions for which $F(x) \to 0$ as $x \to -\infty$ and $F(x) \to 1$ as $x \to \infty$.

The notion of random variable and c.d.f. extends easily to k-dimensional random variables. A stochastic process $\{X_t, t \in T\}$ concerns infinite-dimensional random variables if T is an infinite set. The distribution of these are determined, by Kolmogorov's extension theorem, by the finite-dimensional distributions.

The *expectation* or *mean* of X is defined by

$$E(X) = \int x \, dF(x),$$

if $\int |x| \, dF(x) < \infty$. We shall extend this definition to include $+\infty$ and $-\infty$ as possible values of $E(X)$ as long as not both $\int_{-\infty}^{0} |x| \, dF(x)$ and $\int_{0}^{\infty} |x| \, dF(x)$ are infinite. This definition is designed to represent the long run average of the observed values of X derived from repetitions of the experiment.

The above definition implies that

$$(1.2) \qquad E(X) = \int X(z) P(dz)$$

$$(1.3) \qquad E(X + Y) = E(X) + E(Y)$$

1

if X and Y are random variables with finite expectations. Moreover, if h is a Borel measurable real-valued function,

$$(1.4) \qquad E[h(X)] = \int h(X(z))P(dz) = \int h(x)dF(x).$$

The *conditional probability* of B given A, $P(B|A)$ is defined by

$$(1.5) \qquad P(A \cap B) = P(A)P(B|A), \qquad\qquad A, B \in \mathscr{B},$$

when $P(A) \neq 0$, and represents (in the frequency interpretation) the long run proportion of times that B will succeed among those trials in which A succeeds. This leads to the notion of independence. Ordinarily one would define B to be independent of A if $P(B|A) = P(B)$. But more conventional is the symmetric definition which does not invoke $P(A) > 0$; the events A and B are *independent* if

$$(1.6) \qquad P(AB) = P(A)P(B).$$

Several events $A_1, A_2, \cdots, A_k$ are *(mutually) independent* if

$$P(A_{i_1} \cdot A_{i_2} \cdots A_{i_r}) = \prod_{j=1}^{r} P(A_{i_j})$$

for every subset of distinct integers, $i_1, i_2, \cdots, i_r$ from 1 to k.

A set of random variables $X_1, X_2, \cdots, X_k$ are independent if the events $(X_i \in A_i)$, $i = 1, 2, \cdots, k$, are independent for all Borel sets $A_1, A_2, \cdots, A_k$. It can be shown that

$$(1.7) \qquad E\left[\prod_{i=1}^{k} h_i(X_i)\right] = \prod_{i=1}^{k} Eh_i(X_i)$$

if the X_i are independent random variables and the $h_i(X_i)$ have finite expectations.

The definition of expectation is "justified" by the *law of large numbers*, which states that if $X_1, X_2, \cdots, X_n, \cdots$ are *independent identically distributed* (i.i.d.) random variables with expectation μ, then

$$(1.8) \qquad \bar{X}_n = \frac{1}{n}(X_1 + X_2 + \cdots + X_n) \to \mu \quad \text{w.p.1},$$

where w.p.1 stands for "with probability 1" or equivalently with the exception of a set of (probability) measure zero.

A measure of variation, is the *standard deviation* σ_X which is the square root of the *variance* defined by

$$(1.9) \qquad \sigma_X^2 = E[(X - \mu_X)^2],$$

where

$$(1.10) \qquad \mu_X = E(X).$$

This concept is generalized to the vector-valued random variable X whose transpose is

$$X' = (X_1, X_2, \cdots, X_n).$$

Then

$$\mu_X' = E(X') = (E(X_1), E(X_2), \cdots, E(X_n)) = (\mu_{X_1}, \mu_{X_2}, \cdots, \mu_{X_n})$$

is the transpose of the mean vector μ_X, and the covariance matrix is defined by

(1.11) $$\Sigma_{XX} = E[(X - \mu_X)(X - \mu_X)'],$$

whose general term is the covariance of X_i and X_j,

(1.12) $$\sigma_{X_i X_j} = E[(X_i - \mu_{X_i})(X_j - \mu_{X_j})].$$

If X_i and X_j are independent, then $\sigma_{X_i X_j} = 0$. If $\sigma_{X_i X_j} = 0$, we say that X_i and X_j are uncorrelated, but this does not imply independence. If all the variables are pairwise uncorrelated, i.e., every pair X_i and X_j are uncorrelated, then Σ_{XX} is a diagonal matrix, and the variance of their sum is the sum of their variances. A consequence of great importance is that

(1.13) $$\sigma_{\bar{X}}^2 = \sigma^2/n$$

when $\bar{X}$ is the average of n independent observations on a random variable X with variance σ^2.

2. Generalities about the conventional theory of design of experiments. Design of experiments received its major impetus from the work of R. A. Fisher and others interested in agricultural and biological applications. Major considerations were the need to measure many quantities, the need to avoid the possibility of serious error because of possible complicating factors difficult to evaluate in advance, and the need for computational simplicity. For example, in comparing several fertilizers in their effect on crop yield, it is not satisfactory to plant one plot (of land) with each fertilizer, for such a procedure would leave the statistician unable to determine whether the better crop was due to the fertilizer or to the local variation in land quality. If one could assume that variations in land fertility were roughly linear, it might be possible to rearrange the area to be planted into many subplots, several of which would be assigned to each fertilizer, in such a way that analysis of resulting data could be used to cancel out the effect of land fertility, measure the desired differences in fertilizer effects, and even to measure how dependable these estimates are.

On the other hand, if land fertilities varied erratically, the device of randomization, where plots were allocated at least partly by a random scheme, could be used effectively to cancel out undesirable "bias". The use of randomization raises a philosophical question which is articulated by the following probably apocryphal anecdote.

The metallurgist told his friend the statistician how he planned to test the effect of heat on the strength of a metal bar by sawing the bar into six pieces. The first two would go into the hot oven, the next two into the medium oven, and the last two into the cool oven. The statistician, horrified, explained how he should randomize to avoid the effect of a possible gradient of strength in the metal bar. The method of randomization was applied, and it turned out that the randomized experiment called for putting the first two pieces into the hot oven, the next two into the medium oven, and the last two into the cool oven. "Obviously, we can't do that," said the metallurgist. "On the contrary, you have to do that," said the statistician.

What are the arguments for and against this design? In a "larger" design or sample, the effect of a reasonable randomization scheme would be such that this obvious difficulty would almost certainly not happen. Assuming that the original strength of the bar and the heat treatment did not "interact" in a complicated nonlinear way, the randomization would virtually cancel out any effect due to a strength gradient or other erratic phenomena, and computing estimates as though these did not exist would lead to no real error. In this small problem, the effect may not be cancelled out, but the statistician still has a right to close his eyes to the design actually selected if he is satisfied with "playing fair". That is, if he instructs an agent to select the design and he analyzes the results, assuming there are no gradients, his conclusions will be *unbiased* in the sense that a tendency to overestimate is balanced on the average by a tendency to underestimate the desired quantities. However, this tendency may be substantial as measured by the variability of the estimates which will be affected by substantial gradients. On the other hand, following the natural inclination to reject an obviously unsatisfactory design resulting from randomization puts the statistician in the position of not "playing fair". What is worse for an *objective* statistician, he has no way of evaluating in advance how good his procedure is if he can change the rules in the middle of the experiment.

The *Bayesian statistician*, who uses subjective probability and must consider all information, is unsatisfied to simply play fair. When randomization leads to the original unsatisfactory design, he is aware of this information and unwilling to accept the design. In general, the religious Bayesian states that no good and only harm can come from randomized experiments. In principle, he is opposed even to random sampling in opinion polling. However, this principle puts him in untenable computational positions, and a pragmatic Bayesian will often ignore what seems useless design information if there are no obvious quirks in a randomly selected sample.

An interesting illustration of conventional ideas on design will indicate how powerful these ideas can be. This example, essentially due to Hotelling [H1], is based on one by Yates [Y1].

It is desired to estimate the weights of eight objects whose weights are θ_i, $1 \leq i \leq 8$. A chemist's scale is used, which provides a reading which is equal to the weight in one pan minus the weight in the other pan, plus a random error with mean 0 and variance σ^2. Hotelling proposes the design represented by the equations:

$$X_1 = \theta_1 + \theta_2 + \theta_3 + \theta_4 + \theta_5 + \theta_6 + \theta_7 + \theta_8 + u_{1,},$$
$$X_2 = \theta_1 + \theta_2 + \theta_3 - \theta_4 - \theta_5 - \theta_6 - \theta_7 + \theta_8 + u_2,$$
$$X_3 = \theta_1 - \theta_2 - \theta_3 + \theta_4 + \theta_5 - \theta_6 - \theta_7 + \theta_8 + u_3,$$
$$X_4 = \theta_1 - \theta_2 - \theta_3 - \theta_4 - \theta_5 + \theta_6 + \theta_7 + \theta_8 + u_4,$$

$$X_5 = -\theta_1 + \theta_2 - \theta_3 + \theta_4 - \theta_5 + \theta_6 - \theta_7 + \theta_8 + u_5,$$

$$X_6 = -\theta_1 + \theta_2 - \theta_3 - \theta_4 + \theta_5 - \theta_6 + \theta_7 + \theta_8 + u_6,$$

$$X_7 = -\theta_1 - \theta_2 + \theta_3 + \theta_4 - \theta_5 - \theta_6 + \theta_7 + \theta_8 + u_7,$$

$$X_8 = -\theta_1 - \theta_2 + \theta_3 - \theta_4 + \theta_5 + \theta_6 - \theta_7 + \theta_8 + u_8.$$

X_i is the observed outcome of the ith weighing, every $+1$ before a θ_j means that the jth object is in the first pan, and a -1 means that it is in the other pan, and u_i is the random error for the ith weighing and is not observed directly. We estimate the θ_j by solving the equations resulting from assuming $u_i = 0$. This gives, for example, the estimate of θ_1,

$$\hat{\theta}_1 = (X_1 + X_2 + X_3 + X_4 - X_5 - X_6 - X_7 - X_8)/8$$

$$= \theta_1 + (u_1 + u_2 + u_3 + u_5 - u_5 - u_6 - u_7 - u_8)/8.$$

Since the u_i are the errors resulting from independent weighings, we assume that they are independent with mean 0 and variance σ^2. Then a straightforward computation, based on the fact that the matrix of coefficients of the θ_i is a multiple of an orthogonal matrix, yields the result that the $\hat{\theta}_j$ have mean θ_j and covariance matrix $(\sigma^2/8)I$. That is, the $\hat{\theta}_j$ are uncorrelated and each estimates θ_j with variance $\sigma^2/8$.

If one had applied 8 weighings to the first object alone, no better result would have been obtained for θ_1, whereas no estimate would have been available for the other θ_j. Thus a naïve design would have required 64 weighings to achieve these results which have been obtained in 8 weighings by this elegant method of combining various factors simultaneously.

If the number of weights and weighings did not fit exactly, this precise balancing of plus and minus ones would not have been possible, and this unusually high efficiency in estimating all the weights simultaneously would not quite have been achieved.

The kind of balancing shown here has several desirable consequences in many design problems. High efficiency in estimating many factors, simplicity in computation and the cancellation of many complicating additive effects, such as fertility gradients in agricultural experiments, are frequently achievable. To derive such designs, one finds number theoretic results using Latin squares and finite geometry useful.

In these lectures, we shall be concerned with a less complex class of problems. Direct optimization in the estimation of relatively few attributes will be required, and complicating phenomena not in the model will be ignored. On the other hand, randomized designs will appear but for reasons other than those considered above.

3. Optimal sample size. Suppose that it is desired to estimate a quantity θ which can be observed with error u. We may observe $X = \theta + u$, where u is not directly observable and has mean 0 and variance σ^2. Given n independent observations on X, say $X_1, X_2, \cdots, X_n$, we may use the *sample mean*

$$\bar{X} = \frac{1}{n}\sum_{i=1}^{n} X_i = \theta + \frac{1}{n}\sum_{i=1}^{n} u_i$$

as an estimate of θ. How should one select the sample size n? We introduce cost considerations. Let $L(\theta, t)$ be the cost of estimating θ by t and let $C(n)$ be the cost of using a sample of size n. In the special case where

$$L(\theta, t) = k(t - \theta)^2$$

and

$$C(n) = cn$$

the expected cost (*risk*) of using $\bar{X}$ based on a sample of size n is

$$R(\theta, n) = kE[(\bar{X} - \theta)^2] + cn$$
$$= k\sigma^2/n + cn,$$

which is minimized (neglecting the integer character of n) by

$$n_0 = (k\sigma^2/c)^{1/2},$$

and this minimal value is

$$R(\theta, n_0) = 2(ck\sigma^2)^{1/2}.$$

This trivial example raises a number of interesting questions. *Where did the cost structure come from?* In the decision theoretic formulation of statistics, an action is evaluated in terms of its costs. A statistical procedure or strategy is evaluated in terms of the probability distribution of the cost (ordinarily random) resulting from its use. In many problems it is reasonable to use a linear cost function

$$C(n) = c_0 + cn$$

for the cost of taking n observations. The constant c_0 can be neglected since it has no effect on the solution to the problem. The squared error loss $L = k(t - \theta)^2$ is a classical one and may be regarded as a reasonable approximation to a more general loss $L(\theta, t)$ if L is assumed to be zero and to have a minimum when $t = \theta$ and to be smooth in t. Then $L(\theta, t) \approx k(\theta)(t - \theta)^2$ for t close to θ if $\partial^2 L(\theta, \theta)/\partial t^2 > 0$. Assuming that $k(\theta)$ does not vary much with θ and that our estimate will tend to be close to θ gives us the squared error loss as an approximation. It is to be hoped that the solution to this approximate problem would be approximately a solution to the "real" problem. In some problems other forms of L become more relevant, for example, $L(\theta, t) = k|\theta - t|$, but squared error is a more widely applicable form whose use has the additional advantage of making variance relevant.

The variance appears as the *expected* loss. If expectation represents long run average, *what reason would we have to use $EL(\theta, \bar{X})$ in a problem which concerns* *us only once?* The answer to this comes from utility theory [L1]. Under very mild assumptions concerning the preferences of an individual, it can be shown that every *prospect* can be expressed in terms of a numerical valued *utility function* which has the following attributes. Of any two prospects, the one preferred by the individual has the higher utility. A lottery which yields two possible prospects with utilities u_1 and u_2 with probabilities p and $1 - p$ respectively, will have utility

$$u = pu_1 + (1 - p)u_2.$$

In short, *the utility of a random prospect is the expectation of the resulting utility.* Thus if our losses are measured in utility terms (rather than dollars), the expectation would be a reasonable way of evaluating the random loss associated with $L(\theta, \bar{X})$, before the experiment is carried out. Generally speaking, if relatively small losses are involved, utility tends to be linear in money and there should be little distortion in adding losses or costs such as $L(\theta, \bar{X})$ and $C(n)$.

Why should one use $\bar{X}$ to estimate θ? We shall see that what constitutes an appropriate estimator depends on our knowledge of the probability distribution of the random error. In some problems where this probability distribution is well specified in terms of the *parameter* θ, one can improve considerably on $\bar{X}$. If the errors are normally distributed, $\bar{X}$ is quite a reasonable estimate. We shall elaborate later on the theory of *maximum-likelihood estimation.* In the meantime, we shall simply accept $\bar{X}$ as a reasonable all-purpose estimator whose mean and variance are well-determined by those of the error.

Finally, the proper choice n_0 of n depends on our knowledge of σ^2. *What should we do if σ^2 is not known?* One approach is to guess at σ^2 and act accordingly. Since $R(\theta, n)$ is rather insensitive to small variations in n about n_0, the cost of guessing may not be large. One could proceed sequentially; that is, after n observations one estimates σ^2 in terms of the observed sample variance

$$s^2 = \frac{1}{n-1} \sum_{i=1}^{n} (X_i - \bar{X})^2.$$

Then, one stops sampling if

$$n \geqq (ks^2/c)^{1/2}.$$

Otherwise another observation is taken and the above procedure is repeated. This approach, proposed by Robbins and evaluated by Robbins [R1] and Starr and Woodroofe [S1], would be expected to yield a loss which is a rather good approximation to the optimal value $R(\theta, n_0)$ when σ^2 is known. Indeed the results obtained for the normally distributed errors indicate a remarkably small cost of ignorance of σ^2, a cost measured by that of at most a bounded number of observations (independent of k, σ^2 and c). Calculations show this cost to be about that of wasting one observation. This result indicates a higher order of efficiency than is generally associated with statistical procedures. We shall elaborate on this problem later.

4. Preliminaries on regression. The term *regression* dates back to Galton, who, studying the heights of fathers and sons, noted that the sons of tall fathers tended to be taller than average but *shorter* than their fathers. This phenomenon was labeled regression to indicate the tendency to return toward the average. It can be readily explained in terms of a model where the heights of both father and son are determined by a common genetic trait plus random deviations. Then a tall father will ordinarily indicate partly tall genetic attributes and partly a positive random deviation. While the son will tend to share the genetic attributes, he will average a zero random deviation. Not only does this explain the stated regression phenomenon, but it will predict correctly that the fathers of tall sons will tend to be closer to average than the sons.

In any case, the study of this phenomenon leads one to consider the average or expectation of one random variable given the value of another. When $E|Y| < \infty$, this concept is expressed by the *conditional expectation* of Y given X, $E(Y|X)$ which is (except on a set of probability 0) the unique function of $X = X(z)$ defined by the equation

$$(4.1) \qquad \int_A E(Y|X)P(dz) = \int_A Y(z)P(dz) \quad \text{for} \quad A \in \mathscr{B}(X),$$

where $\mathscr{B}(X)$, called the Borel field generated by X, represents the Borel field generated by sets of the form $\{X \in B\} = \{z : X(z) \in B\}$ and B is a Borel set.

The reader would be well advised to note that if X and Y are random variables, which assume a finite number of possible values, the above definition reduces to the form

$$E(Y|X) = \sum_j y_j P(Y = y_j | X = x_i) \quad \text{if} \quad X = x_i.$$

The definition of conditional expectation implies, under the assumption that $E|Y| < \infty$,

$$(4.2) \qquad \begin{aligned} E[E(Y|X)] &= EY, \\ E[h(X)Y|X] &= h(X)E(Y|X) \quad \text{w.p.1.} \end{aligned}$$

If X and Y are independent,

$$E(Y|X) = E(Y) \quad \text{w.p.1.}$$

Furthermore $E(Y|X)$ has the linear properties of the ordinary expectation. A glance at the defining equation suggests the generalization to $E(Y|\mathscr{F})$, defined for a Borel field $\mathscr{F} \subset \mathscr{B}$, as the $\mathscr{F}$-measurable function on Z, uniquely defined by

$$(4.1) \qquad \int_A E(Y|\mathscr{F})P(dz) = \int_A Y(z)P(dz) \quad \text{for} \quad A \in \mathscr{F}.$$

This generalization is convenient in martingale theory where conditional expectations are taken with respect to an increasing sequence of fields $\mathscr{F}_n$.

In statistical language, the conditional expectation of Y given X is called the *regression* of Y on X.

Let

(4.3) $$Y = E(Y|X) + u$$

define the *residual* u from the regression. It is easy to see that $E(u|X) = 0$ and $E[uE(Y|X)|X] = E(Y|X)E(u|X) = 0$ and hence $Eu = 0$ and $E[uE(Y|X)] = 0$ from which it follows that u and $E(Y|X)$ are uncorrelated (have covariance 0) and

(4.4) $$\sigma_Y^2 = \sigma_{E(Y|X)}^2 + \sigma_u^2.$$

This decomposition suggests the terminology

Total Variance = Explained Variance + Unexplained Variance,

where the regression is regarded as that part of Y explained by (knowledge of) X.

The regression $E(Y|X)$ is the best approximation or prediction of Y based on X from the least squares point of view of minimizing the (conditional) expected squared residual. A variation of this analysis consists of replacing $E(Y|X)$ by the best linear function of X in the sense of minimizing

$$E[Y - (\alpha + \beta X)]^2.$$

Let $Y^* = \alpha + \beta X$ and $u^* = Y - Y^*$. Then the *optimal linear approximation* Y^* satisfies

(4.5) $$Eu^* = E\{[Y - (\alpha + \beta X)]\} = 0,$$
$$Eu^*X = E\{[Y - (\alpha + \beta X)]X\} = 0,$$

from which it follows that

(4.6) $$\sigma_Y^2 = \sigma_{Y^*}^2 + \sigma_{u^*}^2.$$

That portion of the total variance explained by the best linear approximation is the square of the correlation coefficient ρ_{XY}, i.e.,

(4.7) $$\rho_{XY}^2 = \sigma_{Y^*}^2 / \sigma_Y^2.$$

A little computation shows that

(4.8) $$\beta = \sigma_{XY}/\sigma_X^2,$$
$$\alpha = \mu_Y - \beta \mu_X$$

and that

(4.9) $$\rho_{XY} = \sigma_{XY}/\sigma_X \sigma_Y$$

is consistent with the above equation. Note that when the regression is linear, Y^* and $E(Y|X)$ coincide.

Two distinct experimental situations give rise to essentially the same calculations. In one, the variables (X, Y) have a joint distribution and ρ_{XY}^2 measures what portion of the total variance of Y is accounted for by the best linear predictor based on X. In the second, the experimenter may select specific values of x, say, $x_1, x_2, \cdots, x_n$, and observe corresponding values of a random variable Y whose distribution depends on x. The heights of fathers and sons from a certain population is a case of the first example. Our discussion was based on that setup. The case of the heat treatment where the experimenter selects the temperatures of the oven is an example of the second situation.

Suppose, for the sake of discussion, that

$$Y = \alpha + \beta x + u,$$

where u has mean 0 and constant variance σ^2 independent of x. Then the statistician may use the observed values of (x, Y) (note that the deviation u is not observed directly) to estimate α, β and the variance σ^2. As the selected values of x are spread further apart, the fixed variance σ^2 becomes relatively small compared to the increasing spread of the observed Y's. Then the sample correlation,

$$r_{xY} = \frac{\sum (x_i - \bar{x})(Y_i - \bar{Y})}{[\sum (x_i - \bar{x})^2 \cdot \sum (Y_i - \bar{Y})^2]^{1/2}},$$

which is the *sample analogue* of ρ_{xY} in the first experimental situation, approaches ± 1. Thus the correlation is not especially meaningful in the second situation.

Selecting the x_i to increase $\sum (x_i - \bar{x})^2$ tends to improve the estimate of β. One should take care not to spread the x_i so far apart that the linear model with constant variance is no longer a good fit.

The following estimates have validity for both experimental situations. To be specific, let us consider the following generalization of the linear regression model where the x_i are selected and the residuals are independent with constant variance. Let

(4.10) $$Y_i = \beta_1 x_{i1} + \beta_2 x_{i2} + \cdots + \beta_k x_{ik} + u_i, \qquad 1 \leq i \leq n,$$

where the experimenter observes Y_i for specified values of $x_i = (x_{i1}, x_{i2}, \cdots, x_{ik})$, the β_i are unknown *parameters* and the (unobserved) residuals u_i are independent of one another.

Without a serious attempt at justification, let us estimate the β_i by the method of *least squares* which selects the estimates $\hat{\beta}_i$ which are the values of β_i which minimize

$$Q = \sum_{i=1}^{n} (Y_i - \beta_1 x_{i1} - \beta_2 x_{i2} - \cdots - \beta_k x_{ik})^2.$$

Let

$$y = \begin{pmatrix} Y_1 \\ Y_2 \\ \vdots \\ Y_n \end{pmatrix}, \qquad X = \begin{pmatrix} x_{11} & \cdots & x_{1k} \\ \vdots & & \vdots \\ x_{n1} & \cdots & x_{nk} \end{pmatrix}, \qquad u = \begin{pmatrix} u_1 \\ u_2 \\ \vdots \\ u_n \end{pmatrix}.$$

Then

(4.10')
$$y = X\beta + u,$$
$$Q = (y - X\beta)'(y - X\beta),$$

which is minimized at $\beta = \hat{\beta}$ which satisfies

(4.11)
$$X'(y - X\hat{\beta}) = 0,$$
$$\hat{\beta} = (X'X)^{-1}X'y,$$

assuming $X'X$ is nonsingular. Then

(4.12)
$$\hat{y} = X\hat{\beta} = X(X'X)^{-1}X'y$$

is the *projection* of the vector y on the space spanned by the columns of X. The minimized sum of squares is

(4.13)
$$Q = (y - \hat{y})'(y - \hat{y}) = y'y - \hat{y}'\hat{y} = y'y - y'X(X'X)^{-1}X'y.$$

In the form

(4.13')
$$y'y = \hat{y}'\hat{y} + \hat{u}'\hat{u},$$

where $\hat{u} = y - \hat{y}$ is the vector of the so-called *calculated residuals*, this equation is another version of the decomposition of the variance into explained and unexplained parts.

An alternative notation which is suggestive in relating the sample covariances to the population covariances, consists of using

$$M_{YY} = \frac{1}{n}y'y = \frac{1}{n}\sum_{i=1}^{n} Y_i^2,$$

$$M_{Yx} = \frac{1}{n}(y'X) = \left\| \frac{1}{n}\sum_{i=1}^{n} Y_i x_{ij} \right\|,$$

$$M_{xx} = \frac{1}{n}(X'X) = \left\| \frac{1}{n}\sum_{i=1}^{n} x_{ij}x_{ij'} \right\|,$$

where these *moment* matrices are analogous to $\Sigma_{YY}, \Sigma_{YX}, \Sigma_{XX}$. Then

(4.12')
$$\hat{\beta} = M_{xx}^{-1}M_{xY}$$

and the unexplained average sum of squares

(4.13'')
$$\frac{1}{n}Q = \frac{\hat{u}'\hat{u}}{n} = M_{YY} - M_{Yx}M_{xx}^{-1}M_{xY}.$$

Under our assumptions we can evaluate the expectation and covariance of $\hat{\beta}$ and of predictions of $Y(x)$. We have

(4.14)
$$E\hat{\beta} = \beta,$$
$$\Sigma_{\hat{\beta}\hat{\beta}} = \sigma^2(X'X)^{-1} = \frac{\sigma^2}{n}M_{xx}^{-1}.$$

Using

$$\hat{Y}(x_0) = \hat{\beta}'x_0$$

as the prediction of Y for $x = x_0$, we have

$$E\hat{Y}(x_0) = \sum_j \beta_j x_{0j} = \beta'x_0,$$

(4.15)

$$\sigma^2_{\hat{Y}(x_0)} = \sigma^2 x_0'(X'X)^{-1}x_0 = \frac{\sigma^2}{n}x_0' M_{xx}^{-1} x_0.$$

The error in prediction $Y - \hat{Y}(x_0)$, where Y is the observed value when x_0 is applied, has mean 0 and variance $\sigma^2[1 + x_0' M_{xx}^{-1} x_0/n]$.

To derive these results we merely note that the vector u has mean 0 and co-variance $\Sigma_{uu} = \sigma^2 I$. Thus y has mean $X\beta$ and covariance $\Sigma_{yy} = \Sigma_{uu} = \sigma^2 I$. Then

$$\hat{\beta} = (X'X)^{-1}(X'y),$$

$$E\hat{\beta} = (X'X)^{-1}(X'X\beta) = \beta,$$

$$\Sigma_{\hat{\beta}\hat{\beta}} = (X'X)^{-1}X'\Sigma_{yy}X(X'X)^{-1} = \sigma^2(X'X)^{-1},$$

$$E\hat{Y}(x_0) = E(\hat{\beta}'x_0) = \beta'x_0,$$

$$\sigma^2_{\hat{Y}(x_0)} = x_0'\Sigma_{\hat{\beta}\hat{\beta}}x_0 = \sigma^2 x_0'(X'X)^{-1}x_0,$$

and for Y observed at x_0 (independent of the vector y)

$$E(Y - \hat{Y}(x_0)) = \beta'x_0 - \beta'x_0 = 0,$$

$$\sigma^2_{Y-\hat{Y}(x_0)} = \sigma^2_Y + \sigma^2_{\hat{Y}}(x_0) = \sigma^2[1 + x_0'(X'X)^{-1}x_0].$$

Several final remarks are in order. First, if $X'X$ is singular the least squares estimates of $\hat{\beta}$ are not uniquely defined. On the other hand, the projection of y on the space spanned by the columns of X is unique and meaningful. One method used to deal with such cases is that of pseudo-inverses of matrices [A1], [R2], [B7].

Second, most of the computations on estimation are meaningful in the experi-mental situation where X and Y are both random. One proviso must be kept in mind in that case. The disturbances u must be uncorrelated with the elements of X.

Third, the least squares method of estimation has a number of plausible justifica-tions. One which will be discussed later is the applicability of maximum-likelihood under the normal model. Another, which we shall use, is that if it is desired to estimate

$$\varphi = \sum_{j=1}^{k} a_j \beta_j,$$

where the a_i are specified constants, the minimum variance estimate $\hat{\varphi}$ of φ, from the class of linear functions of the Y_i which are *unbiased* (i.e., $E\hat{\varphi} = \varphi$) is the least squares estimator $\sum a_j \hat{\beta}_j$ (which is sometimes uniquely defined even when the vector $\hat{\beta}$ is not). (See Gauss-Markov theorem [S2].)

Finally, the model $Y_i = \beta_1 + \beta_2 x_i + \beta_3 x_i^2 + u_i$ is covered in our linear framework because the regression is *linear in the coefficients*. One may define $x_{i1} = 1$, $x_{i2} = x_i$ and $x_{i3} = x_i^2$, and our analysis applies directly.

5. Design for linear regression: Elfving's method. We consider a class of problems in experimental design where the objective is rather narrowly focused on estimating a specific function of some unknown parameter.

Consider two problems in the design of experiments.

PROBLEM 5.1. Slope of a straight line. Assuming the linear model

$$Y = \alpha + \beta x + u, \qquad\qquad -1 \leqq x \leqq 1,$$

where u has mean 0 and variance σ^2 independent of x, select $x_1, x_2, \cdots, x_n$ to yield an optimal estimate of the slope β.

PROBLEM 5.2. Drug response. The measured response to a dose level x of a drug has mean $\beta_1 x + \beta_2 x^2$ and variance σ^2 independent of x for a reasonable range of x. The model can be represented by

$$Y = \beta_1 x + \beta_2 x^2 + u, \qquad\qquad 0 \leqq x \leqq c.$$

Select $x_1, x_2, \cdots, x_n$ so as to yield an optimal estimate of the response at a proposed nominal level x_0.

Both of these problems are special cases of the following more general regression design problem.

Given the linear regression model

$$Y = \beta_1 x_1 + \cdots + \beta_k x_k + u,$$

where u has mean 0 and variance σ^2 independent of $\mathbf{x} = (x_1, x_2, \cdots, x_k)'$ for $\mathbf{x}$ in a specified set S, select n points of S, $\mathbf{x}_1, \mathbf{x}_2, \cdots, \mathbf{x}_n$ so as to yield an optimal estimate of

$$(5.1) \qquad\qquad \varphi = \sum_{i=1}^{k} a_i \beta_i = \mathbf{a}'\beta.$$

In the first problem $S = \{(x_1, x_2) : x_1 = 1, x_2 = x, -1 \leqq x \leqq 1\}$, $(\beta_1, \beta_2) = (\alpha, \beta)$ and $(a_1, a_2) = (0, 1)$. In the second problem $S = \{(x_1, x_2) : x_1 = x, x_2 = x^2, 0 \leqq x \leqq c\}$ and $(a_1, a_2) = (x_0, x_0^2)$.

Elfving [E1] derived the following elegant graphical solution to this problem.

Solution. Let S^* be the convex set generated by the points of S and S^-, the reflection of S about the origin. Extend the ray from the origin through the point $\mathbf{a}$. The location $\mathbf{z}$ where the ray penetrates S^* represents the solution in the following sense. If $\mathbf{z}$ is a convex combination of points $\mathbf{x}_i$ of S or $-\mathbf{x}_i$ of S^- with weights w_i, assign nw_i observations to "experimental level" $\mathbf{x}_i$. The variance of the estimate of φ is $\sigma^2 \|\mathbf{a}\|^2 / n \|\mathbf{z}\|^2$.

If the nw_i are not exact integers, this solution, ignoring the discrete nature of the integers, is not an exact solution and slightly underestimates the achievable variance.

We illustrate in Figs. 1 and 2 the solution of the above problems before presenting the derivation.

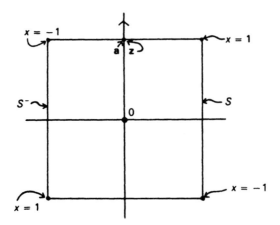

FIG. 1. *Solution for Problem 5.1*

In Problem 5.1 the optimal design consists of using $x = \pm 1$ each half the time. If it were modified so that it were desired to estimate $\alpha + \beta x_0$, then for $x_0 > 1$, one would use $x = +1$ and $x = -1$ in the proportions $(x_0^{-1} + 1)/2$ and $(1 - x_0^{-1})/2$. For $x_0 < 1$ one could use $x = x_0$ for all n observations or $x = +1$ and $x = -1$ in the proportions $(x_0 + 1)/2$ and $(1 - x_0)/2$, among other *designs*. Note that the design where $x = +1$ and -1 in certain proportions has the advantage over the one point design in that it yields estimates of all linear functions of α and β, whereas the one point design provides only an estimate of $\alpha + \beta x_0$.

In Problem 5.2, the optimal design consists of repeating x_0 n times if $(\sqrt{2} - 1)$ $\cdot c < x_0 < c$ or to combine the experimental levels $x = c$ and $x = (\sqrt{2} - 1)c$ in proportions easily determined by the location of the point where the ray from the origin through (x_0, x_0^2) (this ray has slope x_0) penetrates S^*.

The derivation of Elfving's result is instructive. Consider the design which selects, $x_i \in S$, n_i times, $\sum_{i=1}^{m} n_i = n$. The theory of regression suggests that we select the best unbiased linear estimate of the resulting averages $\bar{Y}_i$. Each Y based

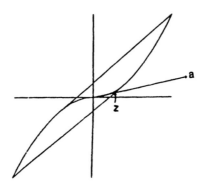

FIG. 2. *Solution for Problem 5.2*

on $\mathbf{x}_i = (x_{i1}, x_{i2}, \cdots, x_{ik})'$ estimates $\beta_1 x_{i1} + \beta_2 x_{i2} + \cdots + \beta_k x_{ik}$ with variance σ^2. It is easy to see that the average $\bar{Y}_i$ is the most efficient linear function of these for estimating $\beta_1 x_{i1} + \beta_2 x_{i2} + \cdots + \beta_k x_{ik}$ and that $\bar{Y}_i$ has variance σ^2/n_i. Thus we confine our attention to estimates of the form

$$(5.2) \qquad \hat{\varphi} = \sum_{i=1}^{m} d_i \bar{Y}_i,$$

where the unbiased condition implies

$$\sum_{i=1}^{m} d_i(\beta_1 x_{i1} + \beta_2 x_{i2} + \cdots + \beta_k x_{ik}) = a_1 \beta_1 + \cdots + a_k \beta_k$$

or

$$(5.3) \qquad \sum_{i=1}^{m} d_i \mathbf{x}_i = \mathbf{a}.$$

Given $\mathbf{d}$ and $\mathbf{x}_1, \mathbf{x}_2, \cdots, \mathbf{x}_m$ we select the n_i to minimize the variance

$$(5.4) \qquad \sigma_{\hat{\varphi}}^2 = \sigma^2 \sum_{i=1}^{m} (d_i^2/n_i).$$

The method of Lagrange multipliers, using the restriction $\sum n_i = n$, yields n_i proportional to $|d_i|$, i.e.,

$$n_i = n|d_i| \Big/ \sum_{j=1}^{m} |d_j|$$

with a resulting value of

$$(5.5) \qquad \sigma_{\hat{\varphi}}^2 = \sigma^2 \left(\sum_{j=1}^{m} |d_j| \right)^2 \Big/ n.$$

Thus we must seek $\mathbf{x}_i$ and d_i subject to (5.3) to minimize $\sum |d_j|$. But (5.3) converts to

$$(5.6) \qquad \sum_{i=1}^{m} w_i[\operatorname{sgn}(d_i)\mathbf{x}_i] = t\mathbf{a},$$

where

$$(5.7) \qquad w_i = |d_i| \Big/ \sum_{j=1}^{m} |d_j|$$

is the proportion of the n observations assigned to $\mathbf{x}_i$ and

$$(5.8) \qquad t = \left[\sum_{j=1}^{m} |d_j| \right]^{-1}.$$

The left side of (5.6) is a point of S^* corresponding to weights w_i attached to points of S or S^- and the right side of (5.6) is a point on the ray from the origin through $\mathbf{a}$. The variance of $\hat{\varphi}$ is minimized by minimizing $\sum |d_j|$ or maximizing t. The Elfving results follow directly.

Several remarks are worth making.

Remark 1. The value of a suboptimal choice of the x_i can be interpreted in terms of the distance from the origin of the point where the convex set generated by these x_i and $-x_i$ is penetrated by the ray from 0 to $\mathbf{a}$. Thus in Problem 5.2, when $(\sqrt{2} - 1)$ $\cdot c < x_0 < c$, the relative value of mixing $x_1 = (\sqrt{2} - 1)c$ and $x_2 = c$ compared to the optimal design of using x_0 exclusively can be measured by considering the ratio of two distances. These are the distance from the origin to (x_0, x_0^2) and the somewhat shorter distance along that ray of slope x_0 to the line segment connecting $((\sqrt{2} - 1)c, (\sqrt{2} - 1)^2c^2)$ with (c, c^2). Thus we see that the cost of using two experimental levels and thereby retaining the ability to estimate response to other drug levels than x_0 is relatively small.

Remark 2. The Elfving result is easily extended in several directions. If the cost per observation depends on x, then a proper formulation would require minimizing $\sigma_{\hat{\phi}}^2$ subject, not to fixed sample size n, but to a fixed total cost K of allocation of x. The results are altered simply by letting w_i represent the proportion of the total cost allocated to the observation at $x = x_i$ whose cost is c_i and replacing S and S^- by $\{x_i c_i^{-1/2}\}$ and $\{-x_i c_i^{-1/2}\}$. The variance of $\sigma_{\hat{\phi}}^2$ is $K^{-1}\sigma^2\|\mathbf{a}\|^2/\|\mathbf{z}\|^2$.

Remark 3. A related extension applies to the case where the variance of u may depend on x. Combining the extension for cost and variance depending on x, we let S be replaced by the set of points $\sigma_i^{-1}c_i^{-1/2}x_i$ for $x_i \in S$, $w_i = \sigma_i c_i^{1/2}|d_i|t$, $t^{-1} = \sum \sigma_i c_i^{1/2}|d_i|$, and $\sigma_{\hat{\phi}}^2 = K^{-1}\|\mathbf{a}\|^2/\|\mathbf{z}\|^2$.

Remark 4. The derivation of Elfving's result depends in an obvious way on the implicit assumption that S is closed and bounded. Granted this assumption, the fact that the solution corresponds to a boundary point of the convex set S^* implies that *there is an optimal design which involves at most k of the available $x \in S$.*

6. Maximum-likelihood estimation. Consider the following two problems.

PROBLEM 6.1. Under a standard environment, the lifetime X of a light bulb is a random variable with probability distribution determined by the c.d.f.

$$F(x|\theta) = P_\theta(X \le x) = 1 - e^{-\theta x}, \qquad\qquad x > 0,$$

where $\theta > 0$ is an unknown parameter (the term *parameter* is used to designate a property of the probability distribution of the data, usually one which determines the probability distribution. Typically, Greek letters are used to represent parameters). It is desired to estimate θ on the basis of n independent observations on X.

PROBLEM 6.2. A random variable X is normally distributed with mean μ and variance σ^2. That is, X has c.d.f.

$$F(x|\theta) = \int_{-\infty}^{x} \frac{1}{\sqrt{2\pi\sigma^2}} e^{-(x'-\mu)^2/2\sigma^2} \, dx',$$

where $\theta = (\mu, \sigma^2)$. On the basis of n observations on X, estimate μ and σ^2.

In both of these problems $F(x|\theta)$ can be represented in terms of a density function with respect to Lebesgue measure. We can represent the c.d.f. by

$$F(x|\theta) = \int_{-\infty}^{x} f(x'|\theta) \, dx',$$

where $f(x'|\theta)$ is called the *density*. Generally we shall deal with families of distributions which have densities with respect to some nonnegative σ-finite measure μ. That is,

$$P_\theta(X \in A) = \int_A f(x|\theta)\mu(dx).$$

In Problem 6.1, the distribution is called *exponential*. It has density $f(x|\theta) = \theta e^{-\theta x}$ for $x > 0$, and mean $\mu_X = E_\theta(X) = \theta^{-1}$ and variance $\sigma_X^2 = \theta^{-2}$.

One approach to estimating θ is to use the *method of moments*. This method, which is not too well specified, consists of estimating a parameter by matching the *sample* moments $m'_r = n^{-1}\sum_{i=1}^n X_i^r$ or the central sample moments $m_r = n^{-1}$ $\cdot \sum_{i=1}^n (X_i - \bar{X})^r$ with the corresponding population moments $\mu'_r = E(X^r)$ or $\mu_r = E(X - \mu_X)^r$. In Problem 6.1, this would give the estimate $\hat{\theta}$ satisfying

$$\hat{\theta}^{-1} = \bar{X} = \frac{1}{n}\sum_{i=1}^n X_i$$

if we match the first moments. If we match the second (central) moments, we would then use

$$\hat{\theta}^{-2} = \frac{1}{n}\sum_{i=1}^n (X_i - \bar{X})^2.$$

While the usual convention is to use the k lowest moments when θ has k components, this illustration raises the point that alternatives exist and choice must be made among them.

Another approach which has desirable large sample properties is that of *maximum-likelihood* estimation. When $f(X|\theta)$ is regarded as a function of θ, this function is called the *likelihood*. The likelihood based on the n independent observations, $X_1, X_2, \cdots, X_n$, is

$$L(\mathbf{X}|\theta) = \prod_{i=1}^n f(X_i|\theta), \qquad \mathbf{X} = (X_1, \cdots, X_n).$$

The maximum likelihood estimator (m.l.e.) is the value $\hat{\theta}$ of θ which maximizes $L(\mathbf{X}|\theta)$. It is easy to see that in Problem 6.1,

$$\hat{\theta} = \bar{X}^{-1},$$

and in Problem 6.2,

$$\hat{\mu} = \bar{X}, \quad \hat{\sigma}^2 = \frac{1}{n}\sum_{i=1}^n (X_i - \bar{X})^2.$$

Any function of the observed data is a *statistic*. Even though a statistic does not involve the unknown value of θ, its probability distribution ordinarily depends on θ. A good estimator T is one which tends to be close to θ irrespective of the value of θ. Hereafter we shall use a circumflex over a parameter to represent the m.l.e. of the parameter unless it is clearly stated otherwise.

What are the properties of the m.l.e.? One large sample (*asymptotic*) property is that $\hat\theta \to \theta$ in probability (*consistency*). Moreover, under regularity conditions $\hat\theta$ is approximately normally distributed with mean θ and "minimal" variance. These terms will be defined carefully, for the asymptotic properties are surprisingly subtle—as a long history of incorrect proofs of almost correct theorems has shown.

To study large sample properties, we consider an estimator t as a sequence of functions $t_n(X_1, X_2, \cdots, X_n)$. For this discussion, let $X_1, X_2, \cdots, X_n, \cdots$ be independent observations on a random variable X with density $f(x|\theta)$ with respect to a measure μ.

To compare various estimators, we note that, using squared error as a measure of the loss involved in estimating θ by $T_n = t_n(X_1, X_2, \cdots, X_n)$, the *risk* or *expected loss* is given by the mean squared error

$$R(\theta, t_n) = E_\theta[L(\theta, T_n)] = E_\theta(T_n - \theta)^2$$

and is ordinarily a function of θ. It is impossible to construct an estimator which is best for all values of θ. For the guess estimator $T_n = \theta_0$ is optimal for $\theta = \theta_0$, but it is relatively poor for other values of θ.

One method used to cope with this situation is to restrict the class of estimators under consideration to *unbiased estimators*, i.e., estimators T_n for which

$$E_\theta(T_n) = \theta \quad \text{for all} \quad \theta.$$

Although that criterion was used in our section on linear regression, it is a questionable one to use in general. In spite of its attractive simplicity, the philosophical foundations for it are weak. Besides, Problem 6.2 furnishes an example which shows one of its weaknesses. Let

$$s_n^2 = \frac{1}{n-1} \sum_{i=1}^n (X_i - \bar{X})^2.$$

Although s_n^2 is used a good deal in practice, partly because $Es_n^2 = \sigma^2$, the estimator $(n-1)s_n^2/(n+1)$ has smaller mean squared error for all σ^2. Furthermore, since s_n^2 is an unbiased estimator of σ^2, s_n *cannot* be an unbiased estimator of σ. Granted that squared error loss in terms of estimates of σ^2 does not convert to squared error loss for σ, one would hope that a principle for restricting the class of rules for estimators under discussion would not depend on the particular choice of the label used to identify the probability distributions of the data.

The advantage of large sample theory in its use in statistics is that as sample sizes grow large, reasonable procedures tend to coincide, and moderately clear-cut rules for (asymptotic) optimality emerge. The main tool for restricting the class of estimators, to eliminate annoying choices like the guess, derives from the observation in Problem 6.2 that as n becomes large, $\bar{X}$ and $\hat\sigma^2$ tend to be very close to μ and σ^2 for each μ and σ^2. Granted that this is a desirable quality, this condition, labeled *consistency*, is invoked in large sample theory. A formal definition follows shortly.

DEFINITION 6.1. A sequence Y_n of random variables *converges in probability* to a random variable Y ($Y_n \to Y$ in prob.) if for each $\varepsilon > 0$,

$$P\{|Y_n - Y| > \varepsilon\} \to 0$$

as $n \to \infty$. If $Y = c$ w.p. 1, we write $Y_n \to c$ in prob.

DEFINITION 6.2. A sequence of distributions P_n *converges in distribution* to P if

(6.1)
$$\int h(z)P_n(dz) \to \int h(z)P(dz)$$

for every continuous bounded function h.

This definition, which requires a topology on the space Z, reduces in the case of real-valued random variables with c.d.f.'s $F_n(x)$ and $F(x)$ to

(6.1')
$$F_n(x) \to F(x) \quad \text{as} \quad n \to \infty$$

for *every point of continuity* of $F(x)$. This is written $F_n \xrightarrow{d} F$ and also applies to the c.d.f. of vector-valued random variables. In that case the c.d.f. is defined by $F(x) = F(x_1, x_2, \cdots, x_k) = P(X_1 \leq x_1, \cdots, X_k \leq x_k)$. If X_n has distribution P_n and X has distribution P we sometimes write

(6.1'')
$$\mathscr{L}(X_n) \to \mathscr{L}(X)$$

to indicate that P_n converges in distribution to P. Mnemonically, $\mathscr{L}$ represents the distribution law.

One should note that since x^2 is not a bounded function, the fact that $F_n \xrightarrow{d} F$ does not imply that $\int x^2 \, dF_n \to \int x^2 \, dF$. In fact, $\liminf \int x^2 \, dF_n \geq \int x^2 \, dF$.

For this reason we shall find it convenient to distinguish between the *limiting variance* and the *asymptotic variance*, i.e., the variance of the limiting distribution.

DEFINITION 6.3. An estimator t is *consistent* if

(6.2)
$$t_n(X_1, X_2, \cdots, X_n) \to \theta \quad \text{in probability.}$$

Under adequate regularity conditions, it follows that the m.l.e. is consistent. A statement and proof of this result is deferred until later. However, it is used to derive the asymptotic distribution of $\sqrt{n}(\hat{\theta}_n - \theta)$.

Let $\theta_0 = (\theta_{10}, \theta_{20}, \cdots, \theta_{k0})'$ represent the *true* value of $\theta = (\theta_1, \theta_2, \cdots, \theta_k)'$, i.e., the value which governs the probability distribution of the independent X_i, $1 \leq i \leq n$. We shall suppress the θ_0 subscript to P and E when there is no danger of ambiguity.

Once consistency has been established, the asymptotic distribution of the m.l.e. becomes relatively easy to find. We shall show that the probability distribution of $\sqrt{n}(\hat{\theta}_n - \theta)$ converges to a normal distribution with mean 0. To do so we use the properties of O_p and o_p, which are the probabilistic analogues of O and o; and the fact that $\mathscr{L}(X_n) \to \mathscr{L}(X)$ implies (a) $\mathscr{L}[g(X_n)] \to \mathscr{L}[g(X)]$ if g is continuous on a set of probability one with respect to $\mathscr{L}(X)$, and (b) $\mathscr{L}(X_n + o_p(1)) \to \mathscr{L}(X)$ if X_n is vector-valued.

Let $Z_n = (X_1, X_2, \cdots, X_n)$. Then, under regularity conditions specified below,

$$\frac{1}{n} \frac{\partial}{\partial \theta} \log L(Z_n|\theta_0) = \frac{1}{n} \frac{\partial}{\partial \theta} \log L(Z_n|\hat{\theta}_n) + \left[\frac{1}{n} \frac{\partial^2}{\partial \theta^2} \log L(Z_n|\theta_0) + o_p(1)\right](\theta_0 - \hat{\theta}_n).$$

Here the $o_p(1)$ term follows from boundedness and continuity conditions as well as consistency. The partial derivative notation is used to represent derivatives of vectors and a matrix with respect to the components of θ. The first term on the right vanishes, under the m.l.e. definition and regularity conditions. As a result,

$$\sqrt{n}(\hat{\theta}_n - \theta_0) = -\left[\frac{1}{n} \frac{\partial^2}{\partial \theta^2} \log L(Z_n|\theta_0) + o_p(1)\right]^{-1} \cdot \left[n^{-1/2} \frac{\partial}{\partial \theta} \log L(Z_n|\theta_0)\right].$$

But

$$\frac{\partial}{\partial \theta} \log L(Z_n|\theta_0) = \sum_{i=1}^{n} \frac{\partial \log f(X_i|\theta_0)}{\partial \theta},$$

which is a sum of i.i.d. random variables with mean

$$E \frac{\partial \log f(X|\theta_0)}{\partial \theta} = \int \frac{\partial f}{\partial \theta}(x|\theta_0)\mu(dx) = \frac{\partial}{\partial \theta} \int f(x|\theta_0)\mu(dx) = \frac{\partial}{\partial \theta}(1) = 0$$

and nonnegative definite covariance matrix

$$J(\theta_0) = E\left\{\left[\frac{\partial \log f(X|\theta_0)}{\partial \theta}\right] \cdot \left[\frac{\partial \log f(X|\theta_0)}{\partial \theta}\right]'\right\}.$$

Thus, the central limit theorem implies that

$$\mathcal{L}\left[n^{-1/2} \frac{\partial}{\partial \theta} \log L(Z_n|\theta_0)\right] \rightarrow N(0, J(\theta_0)),$$

where $N(\mu, \Sigma)$ represents the multivariate normal distribution with mean μ and covariance matrix Σ.[1]
Note that

$$E \frac{\partial^2}{\partial \theta^2} \log f(X|\theta_0) = E\left\{-\left[\frac{\partial \log f(X|\theta_0)}{\partial \theta}\right] \cdot \left[\frac{\partial \log f(X|\theta_0)}{\partial \theta}\right]'\right.$$

$$\left. + \frac{1}{f(X|\theta_0)} \frac{\partial^2 f(X|\theta_0)}{\partial \theta^2}\right\} = -J(\theta_0),$$

---

[1] The standard normal distribution is defined by the density $\varphi(x) = (2\pi)^{-1/2} e^{-x^2/2}$ and has mean 0 and variance 1. Taking i.i.d. random variables $X_1, X_2, \cdots, X_k$ with the standard normal distribution, the r.v. $Y = a + BX$ has mean $\mu_Y = a$, and covariance matrix $\Sigma_{YY} = BB'$. This distribution, labeled $\mathcal{N}(\mu_Y, \Sigma_{YY})$ because it is determined by μ_Y and Σ_{YY}, has density $(2\pi)^{-k/2}(\det \Sigma_{YY})^{-1/2} \exp[-(y - \mu_Y)' \cdot \Sigma_{YY}^{-1}(y - \mu_Y)/2]$ when Σ_{YY} is positive definite.

and since

$$\frac{\partial^2}{\partial\theta^2}\log L(Z_n|\theta_0) = \sum_{i=1}^{n}\frac{\partial^2}{\partial\theta^2}\log f(X_i|\theta_0),$$

the law of large numbers implies that

$$\frac{1}{n}\frac{\partial^2}{\partial\theta^2}\log L(Z_n|\theta_0) = -J(\theta_0) + o_p(1).$$

Thus, if $J(\theta_0)$ is positive definite

$$\sqrt{n}(\hat{\theta}_n - \theta_0) = [J(\theta_0)]^{-1}Y_n + o_p(1),$$

where $\mathcal{L}(Y_n) \to \mathcal{N}(0, J(\theta_0))$ and it follows that

$$\mathcal{L}(\sqrt{n}(\hat{\theta}_n - \theta_0)) \to \mathcal{N}(0, [J(\theta_0)]^{-1}).$$

Stated formally, we have the following theorem, where $|A|$ denotes the maximum absolute value of the elements of a matrix A.

THEOREM 6.1. *If* $\hat{\theta}_n \to \theta_0$ *in prob.,* $E_{\theta_0}[\partial \log f(X|\theta_0)/\partial\theta] = 0$, $E_{\theta_0}[f^{-1}(X|\theta_0) \cdot \partial^2 f(X|\theta_0)/\partial\theta^2] = 0$, $J(\theta_0)$ *is positive definite, and* $E_{\theta_0}W(\delta) \to 0$ *as* $\delta \to 0$ *where* $W(\delta) = \sup_{|\theta-\theta_0|<\delta}|\partial^2 \log f(X|\theta)/\partial\theta^2 - \partial^2 \log f(X|\theta_0)/\partial\theta^2|$, *then*

(6.3)
$$\mathcal{L}(\sqrt{n}(\hat{\theta}_n - \theta_0)) \to \mathcal{N}(0, J^{-1}(\theta_0)).$$

The nonnegative definite matrix

(6.4)
$$J(\theta_0) = E_{\theta_0}\left[\frac{\partial \log f(X|\theta_0)}{\partial\theta}\right] \cdot \left[\frac{\partial \log f(X|\theta_0)}{\partial\theta}\right]'$$

is called *Fisher's information matrix*. Part of the justification for this term lies in the fact that the information matrix based on X and Y, where X and Y are independent, is the sum of the individual matrices, i.e.,

(6.5)
$$J_{(X,Y)}(\theta) = J_X(\theta) + J_Y(\theta).$$

We shall find useful the easily established fact that the randomized experiment where the r.v. X_i is observed with probability w_i has information matrix $\sum w_i J_{X_i}(\theta)$. Basically, the main justification for the term was a long conjectured result that J^{-1} describes the asymptotic covariance of $\hat{\theta}_n$ and that one *could not do better*. One basis for this conjecture is the Cramér–Rao inequality.

Let T_n be an estimate of $\theta = (\theta_1, \theta_2, \cdots, \theta_k)'$. For any vector $a = (a_1, a_2, \cdots, a_k)'$ let

$$V_n(a, \theta) = E_\theta\{[(a'T_n) - (a'\theta)]^2\}$$

be the mean squared deviation of $a'T_n$ from $a'\theta$.

THEOREM 6.2 (Cramér–Rao inequality). *If* T_n *is an unbiased estimate of* θ, *then*

$$V_n(a, \theta_0) \geq \frac{1}{n}a'[J(\theta_0)]^{-1}a$$

under the regularity conditions: (i) $J(\theta_0)$ *is positive definite*, (ii) $E_{\theta_0}[\partial \log f(X|\theta_0)/\partial\theta]$ $= 0$, (iii) $E_\theta[T_n(\partial \log L(Z_n|\theta)/\partial\theta)'] = \partial E_\theta[T_n]/\partial\theta = I$ at $\theta = \theta_0$.

There are many variations of this theorem. The above statement is one of the weaker forms except for the fact that it deals with a k-dimensional parameter. The proof which follows has the basis for minor extensions for biased estimators and singular information matrices, and starts without at first assuming that the bias

$$b_n(\theta) = E_\theta T_n - \theta$$

is zero.

Proof. Let a^* be an arbitrary vector in k-dimensional space. Then by the Schwarz inequality,

$$\left\{E\left(a'(T_n - \theta_0)\left[\frac{\partial \log L(Z_n|\theta_0)}{\partial\theta}\right]'a^*\right)\right\}^2 \leq E\left\{[a'(T_n - \theta_0)(T_n - \theta_0)'a]\right.$$
$$\left. \cdot E\left[a^{*'}\left(\frac{\partial \log L}{\partial\theta}\right)\left(\frac{\partial \log L}{\partial\theta}\right)'a^*\right]\right\}.$$

After interchanging derivative and expectation in the left-hand side, this inequality becomes

$$\left\{a'\left[I + \frac{\partial b_n(\theta_0)}{\partial\theta}\right]a^*\right\}^2 \leq nV_n(a, \theta_0) \cdot a^{*'}J(\theta_0)a^*.$$

Since T_n is unbiased, if we substitute $a^* = [J(\theta_0)]^{-1}a$ we have

$$V_n(a, \theta_0) \geq n^{-1} \cdot a'J(\theta_0)^{-1}a.$$

This result combined with consistency strongly suggests that one cannot estimate θ any better than indicated by the covariance matrix $n^{-1}J^{-1}$. A number of attempts to prove such an optimality result failed, even when attention was confined to asymptotically normal competitors. Basically the reason is that a pointwise limit definition of optimality permitted the phenomenon of *superefficiency* (which is illustrated below), whose presence was not anticipated in the previous attempts.

For the variation of Problem 6.2 where it is known that $\sigma = 1$, and $\hat{\theta}_n = \hat{\mu}_n = \bar{X}$, we let $T_n = \bar{X}$ as long as $|\bar{X}| \geq n^{-1/4}$. Otherwise, let $T_n = \frac{1}{2}\bar{X}$. Then, if $\theta \neq 0$, $\mathcal{L}(\sqrt{n}(T_n - \theta)) \to \mathcal{N}(0, 1)$ and if $\theta = 0$, $\mathcal{L}(\sqrt{n}(T_n - \theta)) \to N(0, 1/4)$. On the other hand, $\mathcal{L}(\sqrt{n}(\bar{X} - \theta)) \to \mathcal{N}(0, 1) = \mathcal{N}(0, J^{-1}(\theta))$ for all θ. Thus, asymptotically, T_n is as good as $\bar{X}$ for all θ and *better* for $\theta = 0$. Actually, for finite n the improvement at $\theta = 0$ is bought at a cost for θ moderately close to 0. As $n \to \infty$, the place of reduced performance moves away from any specified $\theta \neq 0$ toward $\theta = 0$ but never affects $\theta = 0$.

This illustrates a potential danger in all asymptotic criteria of optimality, which should never be forgotten. We use such criteria as a means of suggesting good procedures whose goodness must be checked out in the "real" world of finite n. The above illustration of superefficiency can be extended to sets of measure zero in the space of possible θ.

Let us summarize. Under regularity conditions, the m.l.e. $\hat{\theta}_n$ is consistent and the limiting distribution of $\sqrt{n}(\hat{\theta}_n - \theta)$ is $\mathcal{N}[0, J(\theta)^{-1}]$. To establish optimality, the Cramér–Rao inequality tries to show that *one cannot do better.* Unfortunately superefficiency shows that an appropriate statement of optimality cannot be too simple. Our version of the Cramér–Rao inequality has three defects. It invokes unbiasedness and it has a regularity condition involving the candidate estimator T_n. This condition suggests that there may be a nonregular estimator for which the bound does not hold. Finally it fails to address the possibility of the superefficiency phenomenon.

Another complication arises from the fact that asymptotic variance and limiting variance differ. Our interest in variance derives from the expansion

$$L(\theta, T) \approx c(\theta)(T - \theta)^2 + \cdots.$$

But if $nE(T - \theta)^2$ and the second moment of the limiting distribution of $\sqrt{n}(T - \theta)$ do not coincide (typically the first exceeds the second), which is relevant?

These issues are discussed in [C1]. Commins [C2] and later Joshi [J1] have shown that for the m.l.e. the two coincide and are relevant. Related versions of the Cramér–Rao theorem, which do not invoke unbiasedness nor any other regularity conditions on the estimator, are due to Chapman and Robbins [C3] and Stein (unpublished). This latter version can be applied to the asymptotic case to show that *even over vanishingly small θ intervals, the maximum of the normalized risk $E_\theta[nL(\theta, T)/c(\theta)]$ times the information $J(\theta)$ can not be bounded below* 1. In this form the superefficiency phenomenon is allowed for and we have a reasonable version of the asymptotic optimality for the m.l.e. Furthermore, $J(\theta)$ constitutes a reasonable measure of the information available in X.

Incidentally this version of the Cramér–Rao inequality has the added advantage that if the observations are independent but not identically distributed, the information J can be replaced by the average $\bar{J}_n$ of the n individual informations.

Le Cam's approach to asymptotic optimality [L2] invokes the underlying idea that if one assumes a prior distribution, the *Bayes* solution, i.e., the estimator which minimizes the average loss with respect to this prior distribution, has an asymptotic behavior that is rather insensitive to the prior distribution and to the loss function. Comparing the m.l.e. with suitable Bayes solutions yields optimality properties of the maximum-likelihood estimators. Notice that superefficiency effects tend to be damped out when $E_\theta L(\theta, T)$ is averaged with respect to a prior distribution on θ.

7. Locally optimal designs for estimation.
Consider two problems.

PROBLEM 7.1. The efficacy of a drug at dose level x is determined by the probability of response

$$p(x) = 1 - e^{-\theta x}, \qquad\qquad 0 < x < \infty, \theta > 0.$$

It is desired to estimate θ.

PROBLEM 7.2. To determine how strongly smoking is correlated with heart disease, one may choose among several experiments. One may select a person at random from the population and determine whether or not he smokes and whether or not he has heart disease.

	H	$\tilde{H}$	
S	p_{SH}	$p_{S\tilde{H}}$	p_S
$\tilde{S}$	$p_{\tilde{S}H}$	$p_{\tilde{S}\tilde{H}}$	$p_{\tilde{S}}$
	p_H	$p_{\tilde{H}}$	

The possibilities are covered in the 2×2 table above where the entries represent the appropriate probabilities. Alternatively, one may confine attention to smokers, sampling one at random to determine whether he has heart disease. Or one may sample among nonsmokers or among heart disease victims or non-heart disease victims. In Problem 7.2 we shall assume p_S and p_H are known but leaving essentially one unknown parameter, say,

$$\theta = p_{SH} - p_S p_H,$$

which is to be estimated efficiently.

In Problem 7.1, the density is given by

$$f(y|\theta, x) = e^{-\theta x} \qquad \text{for} \quad y = \text{no response},$$

$$f(y|\theta, x) = 1 - e^{-\theta x} \quad \text{for} \quad y = \text{response}.$$

(This is a density with respect to μ which assigns measure 1 to both possible outcomes. This is the customary counting measure used for discrete-valued random variables.) Then

$$J(\theta) = E_\theta \left(\frac{\partial \log f}{\partial \theta} \right)^2 = x^2 e^{-\theta x} + \frac{x^2 e^{-2\theta x}}{(1 - e^{-\theta x})^2} \cdot (1 - e^{-\theta x})$$

$$= \frac{1}{\theta^2} (\theta x)^2 \frac{e^{-\theta x}}{1 - e^{-\theta x}},$$

which is maximized for $\theta x = 1.6$, i.e., $e^{-\theta x} \approx 0.2$, with a maximum value of $0.64\theta^{-2}$. Thus from the point of view of information, the best experimental design consists of selecting x so that $x = 1.6\theta^{-1}$, and the probability of response is about 0.8.

Of course, if θ is not known, one cannot select x appropriately. However, one may easily imagine that there would be relatively little loss in proceeding sequentially: estimating θ after the nth observation and using that estimate to improve somewhat the value of x to be used next time. Since $J(\theta)$ is relatively flat near its peak, there would be little loss of efficiency resulting from small deviations in x. (In cases where there is little prior knowledge, customary procedure in bioassay involves starting with a sequence of values of x which are quite widely separated, in order to get a rough idea of what θ is.) Thus, our proposed solution is *locally optimal* in the sense that if θ is approximately known, the optimal experiment is selected. It is convenient to think of the experiment designer as knowing θ_0 the *true* value of θ, when he selects the experiment. In this scenario, the analyzer is not allowed to use the possible relation between the designer's knowledge and the experiment selected to help estimate θ. He must only use the data from the experiment.

If we are given the choice of n x's to select, our recommendation is to select $x = 1.6\theta^{-1}$ n times on the ground that it maximizes the total information available. Is this a sufficient basis for the recommendation? The m.l.e. theory tells us that the resulting m.l.e. estimate $\hat{\theta}_n$ will be such that $\mathcal{L}(\sqrt{n}(\hat{\theta}_n - \theta_0)) = \mathcal{N}(0, 1.56\theta^2)$. Moreover, a consequence of the Stein version of the Cramér–Rao theorem is that for any other design, where the total information is no larger than $nJ(\theta_0) = n(0.64\theta^{-2})$, there is no θ interval in which one can do uniformly better from the point of view of asymptotic risk.

For Problem 7.2, the experiment of selecting a subject who has heart disease and determining whether he smokes can easily be shown to have information

$$J_H = (p_{SH}p_{\bar{S}H})^{-1}$$

Similarly,

$$J_{\bar{H}} = (p_{S\bar{H}}p_{\bar{S}\bar{H}})^{-1},$$

$$J_S = (p_{SH}p_{S\bar{H}})^{-1},$$

$$J_{\bar{S}} = (p_{\bar{S}H}p_{\bar{S}\bar{H}})^{-1},$$

while the selection of an individual at random from the population has information

$$J = p_{SH}^{-1} + p_{\bar{S}H}^{-1} + p_{S\bar{H}}^{-1} + p_{\bar{S}\bar{H}}^{-1},$$

which can be expressed as a weighted average of J_H, $J_{\bar{H}}$, etc. Thus the most informative of these five experiments is to select H if p_H is smaller than $p_{\bar{H}}$, p_S, and $p_{\bar{S}}$.

In both of these problems the best design consists of selecting one experiment and repeating it n times. This is characteristic of problems involving experiments where the outcomes depend on only one unknown parameter.

Matters become more complicated when there are several unknown parameters. Suppose we had two parameters θ_1 and θ_2 and it were desired to estimate θ_1. If there were two experiments available, yielding information matrices

$$\begin{Vmatrix} 4 & 1 \\ 1 & 4 \end{Vmatrix} \quad \text{and} \quad \begin{Vmatrix} 4 & -1 \\ -1 & 4 \end{Vmatrix}$$

respectively, the inverses would be

$$\frac{1}{15} \begin{Vmatrix} 4 & -1 \\ -1 & 4 \end{Vmatrix} \quad \text{and} \quad \frac{1}{15} \begin{Vmatrix} 4 & 1 \\ 1 & 4 \end{Vmatrix}.$$

Thus, the asymptotic variance of $\sqrt{n}(\hat{\theta}_1 - \theta_1)$ of $\hat{\theta}_1$ based on n observations on the first experiment would be 4/15. A similar result would apply for the second. However, combining $n/2$ observations on each of these would yield a total information matrix of $4nI$ with a resulting asymptotic variance for $\sqrt{n}(\hat{\theta}_1 - \theta_1)$ of 1/4.

Thus combining two equally informative experiments could lead to an improvement over using either one of these alone.

To study this problem in greater generality, let us introduce an extension of our terminology. Let $\mathscr{E}$ be a set of available *elementary experiments*. Given an experiment $e \in \mathscr{E}$, we observe a random variable Y_e whose probability distribution depends on e and an unknown vector-valued parameter θ according to a density $f(y|\theta, e)$ with respect to a measure μ_e. A design D consists of a set of n elementary experiments $e_i \in \mathscr{E}$, $1 \leq i \leq n$, (with possible repetitions) to be performed independently. Let

$$(7.1) \qquad J_e(\theta) = E_\theta \left(\frac{\partial \log f(Y_e|\theta, e)}{\partial \theta} \right) \cdot \left(\frac{\partial \log f(Y_e|\theta, e)}{\partial \theta} \right)'$$

be the information matrix corresponding to e.

In view of our previous discussion, it makes sense to relate the problem of selecting an optimal design for estimating θ_1, the first component of θ, with that of selecting $e_1, e_2, \cdots, e_n$ to minimize the $(1, 1)$ element of the inverse of the information,

$$[J_{e_1}(\theta_0) + J_{e_2}(\theta_0) + \cdots + J_{e_n}(\theta_0)]^{11}$$

when θ is believed to be close to θ_0. To minimize that expression is equivalent to minimizing

$$\left[\frac{1}{n} \sum_{i=1}^{n} J_{e_i}(\theta_0) \right]^{11},$$

which corresponds to the randomized experiment which selects e_i with probability $1/n$. It makes sense to extend $\mathscr{E}$ to the class $\mathscr{E}^*$ of randomized experiments. In that case, any design D is equivalent to repeating one randomized experiment n times. Thus we seek $e \in \mathscr{E}^*$ *to minimize*

$$(7.2) \qquad\qquad\qquad [J_e(\theta_0)]^{11}$$

and that experiment is associated with the optimal design for estimating θ_1.

If it were desired to estimate some arbitrary function $\varphi = g(\theta)$, then a transformation of coordinates would convert this problem to the one presented. We shall deal later with the more complex case where one is concerned with several coordinates of θ or several functions of θ.

In the meantime a rather interesting result derives almost immediately from our formulation. We identify with each information matrix $J_e(\theta_0)$, $e \in \mathscr{E}^*$, a point in $(k(k + 1)/2)$-dimensional space representing the elements on and above the main diagonal of J_e. The set S of such points is convex. Assume that it is closed and bounded. Since J^{11} is a "monotone" function of J, the optimum must occur at a boundary point of S in the $(k(k + 1)/2)$-dimensional space and that optimum may be expressed as a convex combination of at most $k(k + 1)/2$ of the points generating S.

Thus, to estimate one parameter using experiments involving at most one other parameter, it is sufficient to use a randomized combination of three elementary experiments. In many applications this reduction serves to reduce the problem of optimal design to simply maximizing a function of three or five variables (the five represents the 3 e_i and the 3 weights subject to the sum equal to 1).

In the regression problem of Elfving, only k experiments are required. Can we do better than $k(k + 1)/2$ here? The answer is, yes. We shall show that in general there is a combination of at most k experiments that suffice for an optimal design for the estimation of θ_1.

THEOREM 7.1. *If S is closed and bounded and J_e is nonsingular for all $e \in \mathscr{E}^*$, then there is a locally optimal design which repeats n times a randomized experiment which is a mixture of at most k of the elementary experiments of $\mathscr{E}$.*

Proof. We expand (suppressing the argument θ_0 in J) for symmetric Δ,

$$(7.3) \qquad (J + h\Delta)^{-1} = J^{-1} - hJ^{-1}\Delta J^{-1} + h^2 J^{-1}\Delta(J + h\Delta)^{-1}\Delta J^{-1},$$

which is valid as long as h is small enough so that $J + h\Delta$ is nonsingular. If the minimum J^{11} is attained at J_0, we express any $J \in S$ as $J = J_0 + h\Delta$, $h > 0$. Then convexity of S implies that $J_0 + h\Delta \in S$ for h sufficiently small and

$$(J_0^{-1}\Delta J_0^{-1})_{11} \leqq 0.$$

Thus J_0 is a value of $J \in S$ which maximizes the linear function $(J_0^{-1}JJ_0^{-1})_{11}$, i.e., J_0 lies on a supporting hyperplane H_1 of S. Moreover, $(J + h\Delta)^{11}$ is constant along the hyperplane $H_2 \subset H_1$ determined by

$$\sum_{i=1}^{k} J_0^{1i}\Delta_{ij} = 0, \qquad\qquad j = 1, 2, \cdots, k.$$

Thus any point of $S \cap H_2$ is an optimal point where H_2 is a $([k(k + 1)/2] - k)$-dimensional hyperplane. But an extreme point of $S \cap H_2$ can be expressed as a convex combination of a set of at most k of the extreme points of S. These correspond to elementary experiments.

Note that the expansion (7.3) implies that for $J \in S$, but not in H_2, $J^{11} > J_0^{11}$. Thus J is optimal if and only if $J \in S \cap H_2$.

Two major points require discussion. These are the possibility of singular information matrices and the case where several parameters are to be estimated.

First, let us discuss several relatively minor issues. The optimal design which consists of repeating one randomized experiment n times would offend a Bayesian. In the context of our problem, the randomization serves a much simpler function than in the more conventional design problems. Here it is simply a mechanism of saying that we desire our elementary experiments to be allocated in certain proportions and our goal can be achieved by a deterministic allocation of experiments. If we chose to allocate between two experiments in proportions $2/\pi$ and $1 - 2/\pi$, then for any fixed integer n, a nonrandomized design would not achieve the goal. Thus the militant nonrandomizer requires a supplement to the main theorem which announces that J^{11} is continuous in J at the solution J_0. This supplement assures that for proportions close to the optimal ones, the relevant variance is close to the optimal. This continuity result is trivial under the nonsingular restriction.

The extension of Theorem 7.1 from the case of constant cost per experiment and specified n, to variable cost and specified total cost is easy. One simply measures information per unit cost for each e and the proportions allocated in the optimal design refer to proportions of total cost.

Suppose that it is desired to estimate r functions of the k parameters. In nondegenerate cases, we may reparametrize θ so that we are interested in the first r components of θ. Using a quadratic loss function,

$$L(\theta, T) = \sum_{i,j=1}^{r} a_{ij}(T_i - \theta_i)(T_j - \theta_j) = (T - \theta)'A(T - \theta),$$

the estimate t has risk

(7.4) $R(\theta, t) = E_\theta(T - \theta)'A(T - \theta) = \text{tr}(A\Sigma_{TT})$,

where

(7.5) $\Sigma_{TT} = E_\theta[(T - \theta)(T - \theta)']$

and A has rank no greater than r. If A had rank less than r, we could reduce the problem (locally) to estimating fewer than r parameters. Thus we may assume that A has rank r and, by use of a linear transformation, that the components of A vanish except for ones for the first r diagonal elements. In analogue to the one-dimensional case, we seek therefore to minimize the sum of the asymptotic variances of the estimates of the first r components of θ, or to select the element of S which minimizes

(7.6) $\text{tr}(AJ^{-1}) = J^{11} + J^{22} + \cdots + J^{rr}$.

THEOREM 7.2. *If S is closed and bounded and J is nonsingular for all $J \in S$, $\sum_{i=1}^{r} J^{ii}$ is minimized for $J \in S$ by a convex combination of at most $k + (k - 1) + \cdots + (k - r + 1) = r(2k - r + 1)/2$ of the extreme points of S.*

Proof. Applying (7.3) to $J_0 + h\Delta \in S$ with J_0 optimal and $h > 0$, we have

$$\sum_{m=1}^{r} \sum_{i,j} J_0^{mi} \Delta_{ij} J_0^{jm} \leq 0,$$

i.e., J_0 is on a supporting hyperplane of S, H_3. Moreover, the submatrix $\|J^{ij}\|$, $1 \leq i, j \leq r$, is constant on the hyperplane $H_4 \subset H_3$ defined by

$$\sum_{j=1}^{k} \Delta_{ij} J_0^{jm} = 0, \qquad i = 1, 2, \cdots, k, m = 1, 2, \cdots, r.$$

Because of the symmetry of Δ_{ij}, H_4 is determined by $k + (k - 1) + \cdots + (k - r + 1)$ linear restrictions. Thus there is an extreme point of $H_4 \cap S$ which is a convex combination of $r(2k - r + 1)/2$ extreme points of S and which minimizes $\sum_{i=1}^{r} J^{ii}$.

This proof applies equally well to any function of the submatrix $\|J^{ij}\|$, $1 \leq i$, $j \leq r$, which cannot attain a minimum at an interior point of an available set.

Up to now we have neglected the problem which exists when the information matrix can be singular. This is a serious problem because of many important applications. For example, in the regression problem treated by Elfving, an elementary experiment corresponds to the observation of

$$Y(x) = \beta_1 x_1 + \beta_2 x_2 + \cdots + \beta_k x_k + u.$$

If we assume that u is normally distributed with mean 0 and variance 1, the corresponding information matrix becomes

$$J(x) = xx'$$

which has *rank one*. But this is precisely the case whenever an elementary experiment yields information on only one function of the unknown parameter.[2] The following probit analysis problem used in bioassay and reliability is a nonregression illustration.

PROBLEM 7.3. We assume that the dose level of a drug required to achieve a response in the given population is normally distributed with unknown mean μ and variance σ^2. Thus the probability of response to a dose d is

$$p(\theta, d) = \int_{-\infty}^{d} \frac{1}{\sqrt{2\pi\sigma^2}} e^{-(x-\mu)^2/2\sigma^2} \, dx = \Phi\left(\frac{d - \mu}{\sigma}\right),$$

where

$$\Phi(x) = \int_{-\infty}^{x} (2\pi)^{-1/2} e^{-t^2/2} \, dt = \int_{-\infty}^{x} \varphi(t) \, dt$$

is the standard normal c.d.f. It is desired to select levels $d_1, d_2, \cdots, d_n$ so as to minimize the variance of the estimate of $\psi = \mu - 2.87\sigma$, the dose level required to achieve probability 0.002 of response.

For this problem the information matrix corresponding to the dose level d becomes

$$J_d(\theta) = \frac{\varphi^2}{\sigma^2 \Phi(1 - \Phi)} \left\| \begin{matrix} 1 & x_d \\ x_d & x_d^2 \end{matrix} \right\|,$$

where φ and Φ are evaluated at $x_d = (d - \mu)/\sigma$.

Because a solution of this design problem depends only on the information matrices, and the information matrices for this problem correspond to those of a regression problem with x_1 and x_2 replaced by $\varphi[\Phi(1 - \Phi)\sigma^2]^{-1/2}$ and $x\varphi[\Phi(1 - \Phi)\sigma^2]^{-1/2}$ respectively and a is replaced by $(1, -2.87)$, we can simply apply the Elfving solution to this problem. This solution, represented by Fig. 3,

---

[2] If σ^2 were unknown, then the information matrix would have an extra row and column. Except for the new diagonal element, these would consist of zeros. (This is related to the fact that the m.l.e. of σ^2 is independent of that of the other parameters.) Thus in those regression problems where σ is not a parameter of special interest, it can be regarded as known for the purpose of finding optimal designs.

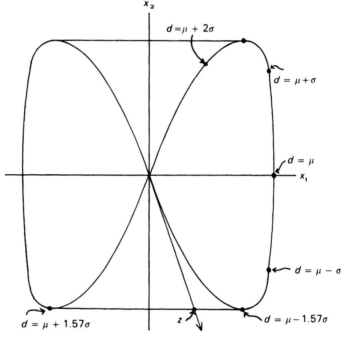

FIG. 3

allocates all of the observations to the two levels $d = \mu \pm 1.57\sigma$ in proportions 0.23 and 0.77.

In this application we have effectively side-stepped the problem of singular information matrices for problems where the elementary experiments have rank 1, by applying Elfving's result for regression problems.

In general, this result is not sufficient. In typical applications there are two basic reasons for singular matrices. One is due to a local singularity illustrated in the problem of estimating $\theta = \mu^{1/3}$ for observations from $\mathcal{N}(\mu, 1)$. Then $J(\theta) = 9\theta^4$ which vanishes at $\theta = 0$. A stretching of the μ-axis or more generally a slight reparametrization of the parameters is helpful in removing such singularities. The other case is where the available data are insufficient to yield consistent estimates of all of the parameters. A trivial illustration is that where one observes Y_e which has the $\mathcal{N}(e(\theta_1 - \theta_2), 1)$ distribution with information matrix

$$J_e = e^2 \left\| \begin{matrix} 1 & -1 \\ -1 & 1 \end{matrix} \right\|,$$

and the available experiments cannot yield estimates of θ_1 or θ_2 separately. A transformation to the parameters $\varphi_1 = \theta_1 - \theta_2, \varphi_2 = \theta_1 + \theta_2$ yields an information matrix

$$J_e^* = e^2 \left\| \begin{matrix} 1 & 0 \\ 0 & 0 \end{matrix} \right\|.$$

We introduce the following variation of the pseudo-inverse to handle the problem of singularity. If A is symmetric and nonnegative definite, let

(7.7)
$$A^{-1} = \lim_{\lambda \to 0+} (A + \lambda B)^{-1},$$

where $A + B$ is positive definite. According to this definition,

$$J_e^{-1} = e^{-2} \begin{Vmatrix} \infty & \infty \\ \infty & \infty \end{Vmatrix},$$

and

$$(J_e^*)^{-1} = e^{-2} \begin{Vmatrix} 1 & a^* \\ a^* & \infty \end{Vmatrix},$$

where a^* depends on B. In general, the elements on the diagonal of A^{-1} are finite or infinite. The (i, j) elements of A^{-1} are finite and independent of B as long as the ith and jth diagonal elements are finite.

This definition is appropriate in the following sense. If $\varphi_1, \varphi_2, \cdots, \varphi_r$ are r independent functions of θ which may be consistently estimated from the data with information matrix $J(\theta)$, there is a locally $1 - 1$ transformation

$$\varphi = h(\theta)$$

with nonsingular Jacobian $\partial \varphi / \partial \theta$ such that the information matrix with respect to φ is

(7.8)
$$J^*(\varphi) = \left(\frac{\partial \theta}{\partial \varphi} \right)' J(\theta) \left(\frac{\partial \theta}{\partial \varphi} \right)$$

and for which the submatrix corresponding to the finite diagonal elements of $[J^*(\varphi)]^{-1}$ measure the appropriate asymptotic covariances.

What is called for is an extension of Theorem 7.3 to the case where J can be singular. In addition, since J^{-1} is not continuous at points of singularity, a Bayesian who wishes to avoid randomized experiments must be shown that approximations to the optimal J using weighted averages of extreme points yield approximately optimal values of $J^{11} + J^{22} + \cdots + J^{rr}$. These results are established in [C4].

8. **More design in regression experiments.** Another generalization of Elfving's work which is applicable to linear regression problems is due to Kiefer and Wolfowitz [K1]. Two alternative criteria of optimality are shown to be equivalent for designs for estimating the coefficients of a linear regression. Reviewing, we have a *closed bounded set S* in k-dimensional space with

$$Y(x) = \beta'x + u, \qquad\qquad x \in S,$$

where u has mean 0 and variance σ^2. Two criteria applied to linear unbiased estimates $\hat{\beta}$ of β are:

I. CHEBYSHEV CRITERION. Select a design to minimize the maximum variance of prediction (among linear unbiased estimators) for $x \in S$, i.e., to

$$\text{minimize} \quad \max_{x \in S} d(x),$$

where

(8.1) $$d(x) = x'\Sigma_{\hat{\beta}\hat{\beta}}x$$

is the variance of the estimate $\hat{\beta}'x$ of $Y(x)$.

II. D-OPTIMALITY. Select a design to

$$\text{minimize} \quad |\Sigma_{\hat{\beta}\hat{\beta}}| = \det(\Sigma_{\hat{\beta}\hat{\beta}}).$$

We shall present a main result which shows that these criteria are equivalent, as well as some extensions and criticisms.

A design which assigns a probability measure μ to S, provides an information matrix per observation

(8.2) $$J_\mu = \int_S (xx')\mu(dx).$$

Then $\Sigma_{\hat{\beta}\hat{\beta}} = n^{-1}\sigma^2 J_\mu^{-1}$ and $nd(x)/\sigma^2$ transforms to

(8.3) $$d(x, \mu) = x'J_\mu^{-1}x.$$

THEOREM 8.1. *Let S be a compact set in E_k and $\sup_\mu |J_\mu| > 0$. Then the conditions*:

  (i) *μ_0 minimizes $|J_\mu^{-1}|$,*
  (ii) *μ_0 minimizes $\sup_{x \in S} d(x, \mu)$,*
  (iii) *$\sup_{x \in S} d(x, \mu_0) = k$,*
define the same set of μ_0. The matrix J_{μ_0} is constant on this set of μ_0.

Before proving Theorem 8.1 (as in [K2]), we shall digress to some necessary background on matrices and game theory. The game theoretic considerations which provide a relatively simple proof are natural in problems involving the minimization of maxima.

8.1. Game theoretic considerations. Let $f(x, y)$ be the payoff to Player I, who selects strategy $x \in X$, from Player II, who selects $y \in Y$. The game defined by (f, X, Y) is said to have a *value v* if

(8.1.1) $$\sup_x \inf_y f(x, y) = \inf_y \sup_x f(x, y) = v.$$

The reason for this term is that in this case Player I can select a *strategy x* which comes arbitrarily close to assuring him of a payoff of v, no matter what II does, even if I announces his strategy. At the same time, II can prevent I from getting a significant amount more than v, no matter what I does.

The strategy x_0 is *conservative for* I if it maximizes $\inf_y f(x, y)$. Similarly, y_0 is *conservative for* II if it minimizes $\sup_x f(x, y)$.

Even in finite games where I and II each have a finite number of available strategies, equation (8.1.1) need not apply. In general we have

$$\inf_y f(x, y) \leq f(x, y'),$$

$$\sup_x \inf_y f(x, y) \leq \sup_x f(x, y')$$

and

(8.1.2)
$$\sup_x \inf_y f(x, y) \leq \inf_y \sup_x f(x, y).$$

The pair of strategies (x_0, y_0) is a *saddle point* of the game if

$$f(x, y_0) \leq f(x_0, y_0) \leq f(x_0, y) \quad \text{for all} \quad x \in X, y \in Y.$$

If the game (f, X, Y) has a saddle point (x_0, y_0), then the game has the value $v = f(x_0, y_0)$, and x_0 and y_0 are conservative strategies for I and II. If the game has conservative strategies x_0 and y_0 for I and II, and the game has a value v, then (x_0, y_0) is a saddle point and $v = f(x_0, y_0)$.

Games involving finite sets X and Y or a continuous function f on a compact set $X \times Y$ always have conservative strategies but do not necessarily have a value and saddle point as the game of matching pennies illustrated below shows.

$$f(x, y)$$

	H	T
H	1	−1
T	−1	1

The fundamental theorem of game theory is *that a finite game, extended to include randomized strategies, has a value and saddle point.* More precisely, we extend (f, X, Y) *to* $(f^*, \Xi, \mathcal{Y})$, *where*

(8.1.3)
$$f^*(\xi, \eta) = \int_{X \times Y} f(x, y)\xi(dx)\eta(dy), \qquad \xi \in \Xi, \eta \in \mathcal{Y},$$

and Ξ and $\mathcal{Y}$ are the class of probability measures on X and Y. The integral above is an expectation which could be more simply written

$$f^*(\xi, \eta) = \sum_{x,y} f(x, y)\xi_x \eta_y, \quad \sum_x \xi_x = 1, \quad \sum_y \eta_y = 1, \xi_x \geq 0, \eta_y \geq 0,$$

in the case of finite X and Y.

One way of approaching this theorem which is instructive is to look at $2 \times n$ games where $X = (1, 2)$ and $Y = (1, 2, \cdots, n)$ and $\xi_2 = 1 - \xi_1 = \xi$. Then every

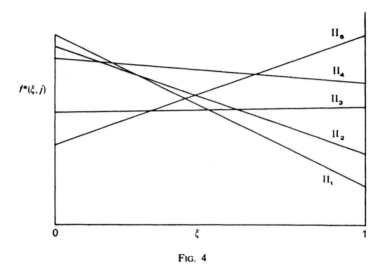

$$f^*(\xi, j)$$

FIG. 4

pure (nonrandomized) strategy of Y is represented by an expected payoff

$$f^*(\xi, j) = (1 - \xi)f(1, j) + \xi f(2, j), \qquad 0 < \xi < 1,$$

which is a line segment in Fig. 4 going from $[0, f(1, j)]$ to $[1, f(2, j)]$. For each ξ, Player II minimizes with respect to η. But

$$\inf_\eta f^*(\xi, \eta) = \min_j f^*(\xi, j)$$

is the minimum of a set of straight lines and is a *concave* function. Indeed *the infimum of a set of concave functions on a convex set S is a concave function* since the condition that for $0 \le \lambda \le 1$, and all $x_1, x_2 \in S$,

$$g_i(\lambda x_1 + (1 - \lambda)x_2) \ge \lambda g_i(x_1) + (1 - \lambda)g_i(x_2) \quad \text{for all } i$$

implies

$$g_i(\lambda x_1 + (1 - \lambda)x_2) \ge \inf_i [\lambda g_i(x_1) + (1 - \lambda)g_i(x_2)] \quad \text{for all } i,$$

$$g_i(\lambda x_1 + (1 - \lambda)x_2) \ge \lambda \inf_i g_i(x_1) + (1 - \lambda) \inf_i g_i(x_2) \quad \text{for all } i,$$

which implies

$$\inf_i g_i(\lambda x_1 + (1 - \lambda)x_2) \ge \lambda \inf_i g_i(x_1) + (1 - \lambda) \inf_i g_i(x_2),$$

which states that $\inf_i g_i$ is concave.

The value ξ_0 of ξ which maximizes $\min_j f^*(\xi, j)$ is Player I's conservative strategy, and looking at the values of j which minimize $f^*(\xi_0, j)$, we obtain a saddle point as follows.

If ξ_0 is not at an endpoint, either there is one minimizing j which corresponds to a horizontal line, or there are at least two minimizing j's with slopes of different signs.

The weighted average of these slopes which provides a slope of zero also provides a randomized strategy for II which prevents I from doing better than $\min_i f^*(\xi_0, i)$, to which he is entitled. These strategies provide a saddle point. If ξ_0 is at an end-point, a little extra argument is necessary.

This point of view is not the simplest for proving the fundamental theorem formally, but it has the advantage of exhibiting the basic mechanism for games $f(x, y)$ which are concave in x or convex in y.

In general, if $f(x, y)$ is continuous on compact $X \times Y$, the extended game (with randomized strategies) has a solution and a saddle point. If $f(x, y)$ is concave in x on a compact convex set X or convex in y on a compact convex set Y, the extended game has a value, and the conservative strategies for the two players form a saddle point. If f is concave in x (convex in y), Player I (II) has a nonrandomized conservative strategy. If f is concave in x and convex in y on compact $X \times Y$, the unextended game has a value and a saddle point.

Incidentally, it is common terminology to call the elements (x_0, y_0) of a saddle point *optimal strategies* for I and II.

8.2. Some properties of determinants.

I. *If A is a nonnegative definite symmetric matrix, then $\operatorname{tr}(AB^{-1})$ is a convex function of B on the set of positive definite symmetric matrices.*

Proof. Let $A = R'R$, where R is a square matrix. Then $\operatorname{tr}(AB^{-1}) = \operatorname{tr}(RB^{-1}R')$. Expanding about $(B + h\Delta)$ we have

$$\operatorname{tr}(R(B + h\Delta)^{-1}R') = \operatorname{tr}[RB^{-1}R' - hRB^{-1}\Delta B^{-1}R' + h^2 RB^{-1}\Delta B^{-1}\Delta B^{-1}R' - \cdots].$$

If $RB^{-1}\Delta \neq 0$, the second order term is positive because B^{-1} is positive definite. If $RB^{-1}\Delta = 0$, the right hand side is $tr(RB^{-1}R')$. Thus we have convexity.

II. *If A is a nonnegative definite $(k \times k)$ symmetric matrix and B is a positive definite $(k \times k)$ symmetric matrix, then*

$$(8.2.1) \qquad \frac{1}{k} \operatorname{tr} AB^{-1} \geq [|A|/|B|]^{1/k}$$

with equality only if A is a scalar multiple of B.

Proof. We may simultaneously diagonalize A and B by the transformation

$$A = RA^*R',$$

$$B = RB^*R',$$

where A^* and B^* are diagonal matrices with diagonal elements $a_i \geq 0$ and $b_i > 0$ and R is a nonsingular matrix. Then $k^{-1} \operatorname{tr}(AB^{-1})$ is the arithmetic average of the a_i/b_i and $[|A|/|B|]^{1/k}$ is the geometric mean of the a_i/b_i. The inequality (8.2.1) follows from the well-known inequality between the arithmetic and geometric means or, by taking logarithms, from the strict concavity of $\log x$. Equality implies that the a_i/b_i are constant, which implies A is a scalar multiple of B.

8.3. Proof of Theorem 8.1. We begin by remarking that $d(x, \mu)$ is not defined if $|J_\mu| = 0$ and hence no such μ satisfies any of the three conditions (i), (ii), (iii). Let us consider the game whose payoff is

$$(8.3.1) \qquad\qquad K(\xi, \mu) = \mathrm{tr}(J_\xi J_\mu^{-1}), \qquad\qquad \xi \in \Xi_\varepsilon, \mu \in \Xi_\varepsilon,$$

where ξ is selected by Player I to maximize and μ by Player II to minimize and $\Xi_\varepsilon = \{\xi : \text{all characteristic roots of } J_\xi \geq \varepsilon\}$.

Since S is compact, the largest characteristic root of J_μ is bounded (say by L) and $|J_\mu|$ attains its maximum M which is positive by assumption. Then for $0 < \varepsilon < ML^{-(k-1)}$, Ξ_ε contains every μ for which $|J_\mu| = M$. Select such an ε. The set Ξ_ε is compact and convex, and K is linear (and hence concave) in ξ and convex in μ. It follows that the game has a saddle point and a value. To prove Theorem 8.1, it will suffice to show that each of the conditions (i), (ii) and (iii) characterize the optimal strategies for Player II and that J_μ is constant over this set of optimal strategies.

Since

$$\inf_\mu K(\xi, \mu) \leq K(\xi, \xi) = k,$$

$$(8.3.2) \qquad\qquad \sup_\xi \inf_\mu K(\xi, \mu) \leq k$$

but

$$(8.3.3) \qquad\qquad \mathrm{tr}\, J_\xi J_\mu^{-1} \geq k[|J_\xi|/|J_\mu|]^{1/k}$$

with strict inequality unless J_ξ is a scalar multiple of J_μ. Hence if ξ is selected to maximize $|J_\xi|$, $K(\xi, \mu) \geq k$ and

$$(8.3.4) \qquad\qquad \sup_\xi \inf_\mu K(\xi, \mu) \geq k.$$

Thus the value of the game is k.

Let $\mathcal{Y}_0$ be the class of optimal strategies for Player II. Then for $\mu_0 \in \mathcal{Y}_0$,

$$(8.3.5) \qquad k \geq K(\xi, \mu_0) = \mathrm{tr}\, J_\xi J_{\mu_0}^{-1} \geq k[|J_\xi|/|J_{\mu_0}|]^{1/k} \quad \text{for all} \quad \xi \in \Xi_\varepsilon.$$

It follows that μ_0 maximizes $|J_\mu|$ and hence $\mathcal{Y}_0 \subset \mathcal{Y}_1 = \{\mu : |J_\mu| \text{ attains its maximum}\}$. Moreover, if one uses any $\xi \in \mathcal{Y}_1$, the resulting equality in (8.3.5) implies that $J_\xi = J_{\mu_0}$. Thus J_μ is constant over $\mathcal{Y}_1$. But $K(\xi, \mu)$ depends on μ only through J_μ. Thus every $\mu \in \mathcal{Y}_1$ is in $\mathcal{Y}_0$ and $\mathcal{Y}_0 = \mathcal{Y}_1$.

Now it remains only to prove that conditions (ii) and (iii) each characterize $\mathcal{Y}_0$. We note that

$$(8.3.6) \qquad k = \sup_{\xi \in \Xi_\varepsilon} K(\xi, \mu_0) < \sup_{\xi \in \Xi_\varepsilon} K(\xi, \mu) \quad \text{for} \quad \mu_0 \in \mathcal{Y}_0, \mu \in \Xi_\varepsilon - \mathcal{Y}_0$$

and

$$(8.3.7) \qquad K(\xi, \mu) = \int_s \mathrm{tr}(xx'J_\mu^{-1})\xi(dx) = \int_s d(x, \mu)\xi(dx) \quad \text{if} \quad |J_\mu| > 0.$$

If Ξ is the set of all probability measures on S, it is easily seen that $\sup_{\xi \in \Xi} K(\xi, \mu)$ $= \max_{x \in S} d(x, \mu)$ by simply assigning ξ probability 1 to the set of x which maximize $d(x, \mu)$. It follows that $\sup_{\xi \in \Xi_\varepsilon} K(\xi, \mu) \to \max_{x \in S} d(x, \mu)$ *monotonically* as $\varepsilon \to 0$. Applying this result to (8.3.6), we see that $\max_{x \in S} d(x, \mu_0) = k$ for every $\mu_0 \in \mathcal{Y}_0$ and $\max_{x \in S} d(x, \mu) > k$ for every $\mu \in \Xi_\varepsilon - \mathcal{Y}_0$ and hence for every μ not in $\mathcal{Y}_0$ for which $|J_\mu| > 0$. But this establishes the desired results.

This theorem which identifies two optimality criteria is the basis for many illustrations developed by Kiefer, Karlin and Studden.

The Chebyshev criterion is a reasonable one when S represents a reasonable range for x and the experimenter wishes to minimize the worst that can happen in his ability to predict over *this range*. It is not useful for extrapolation purposes.

The D-criterion, which consists of minimizing the determinant of the covariance matrix $\Sigma_{\beta\beta}$, and is called the *generalized variance*, has a long history of use in statistics. In my opinion, it is a poor criterion, having no meaningful justification. It abandons to the vagaries of the mathematics of the problem the scientist's function of specifying the loss associated with guessing wrong.

One attractive attribute claimed for the D-criterion is its invariance under nonsingular transformation of parameters. This mathematically neat property simply disguises its shortcomings.

Kiefer [K4] has an extension of the theorem to the case where one is interested in a subset of r of the k parameters. For $r < k$ the analogue of $d(x, \mu)$ is a function whose interpretation reduces the statistical significance of the results.

In the case of estimating r out of k parameters, Kiefer also obtains a result limiting the necessary number of elementary experiments to $r(2k - r + 1)/2$ and for $r = k$ to $k(k + 1)/2$. In a number of examples Kiefer, Karlin and Studden have been able to show that this number is unnecessarily large. Studden [S3] has recently published a more direct extension of Elfving's results to the estimation of several parameters out of k from regression experiments using quadratic loss functions.

9. Testing hypotheses. Statistical inference developed around three major methods. These are estimation, hypothesis testing, and confidence intervals. In the latter one estimates a parameter with an interval which is very likely to contain the parameter. Hypothesis testing principally concerns problems where there are two available actions, although extensions to k-action problems are sometimes considered.

Typical examples involve testing whether or not a treatment has an effect, whether or not a probability distribution fits a particular model, and whether the average quality of a production process exceeds or falls short of a specified level. In each of these examples there are two possible actions. One of these is labeled "accepting the hypothesis" since it typically consists of acting as though the hypothesis were believed to be true, and the other is labeled "rejecting the hypothesis or accepting the alternative hypothesis." In either case, associated with each action is a cost depending on the action and the true situation. (Statisticians typically measure consequences in terms of cost rather than profit.) Data are collected to provide a clue as to what is the true situation or state of nature so that one is more likely to select the best action, i.e., the action which minimizes the cost under the true state of nature.

More formally, we label the two available actions a_1 and a_2 and assume that the state of nature is described by some (unknown) value $\theta \in \Theta$. The cost is given by $L(\theta, a)$. The set Θ may be decomposed into two subsets Θ_1 and Θ_2 so that

$$L(\theta, a_1) \leqq L(\theta, a_2) \quad \text{for} \quad \theta \in \Theta_1,$$

$$L(\theta, a_2) \leqq L(\theta, a_1) \quad \text{for} \quad \theta \in \Theta_2.$$

The action a_i is identified with *accepting the hypothesis* $H_i : \theta \in \Theta_i$, and the two-action problem is called a problem of testing hypotheses.

In the conventional situation where one observes a random variable X with probability distribution P_θ, a *test* or *strategy* s in hypothesis testing is a decomposition of the set X of possible values of X into two sets A_1 and A_2. If $X \in A_1$ the test s leads to *accepting* H_1 (i.e., $s(X) = a_1$ for $X \in A_1$) and if $X \in A_2$, we *reject* H_1 (i.e., accept H_2 or take action a_2). The *error probability* associated with a strategy s is

(9.1)
$$\varepsilon(\theta, s) = P_\theta(A_2) \quad \text{for} \quad \theta \in \Theta_1,$$
$$\varepsilon(\theta, s) = P_\theta(A_1) \quad \text{for} \quad \theta \in \Theta_2.$$

It is customary to replace $L(\theta, a)$ by the regret or loss due to taking the wrong action. Thus L is replaced by L^*, where

$$L^*(\theta, a_1) = 0 \quad \text{and} \quad L^*(\theta, a_2) = L(\theta, a_2) - L(\theta, a_1) \quad \text{for} \quad \theta \in \Theta_1$$

and

$$L^*(\theta, a_2) = 0 \quad \text{and} \quad L^*(\theta, a_1) = L(\theta, a_1) - L(\theta, a_2) \quad \text{for} \quad \theta \in \Theta_2.$$

We introduce the *regret*

(9.2)
$$r(\theta) = |L(\theta, a_1) - L(\theta, a_2)|$$

and the *risk* function is

(9.3)
$$R(\theta, s) = E_\theta L^*[\theta, s(X)] = r(\theta)\varepsilon(\theta, s).$$

The difference between $R(\theta, s)$ and $E_\theta L[\theta, s(X)]$ is $L(\theta, a_i)$ for $\theta \in \Theta_i$ and is not affected by the choice of s. Thus it can be argued that the problem of selecting s to "optimize" R is equivalent to that of "optimizing" $E_\theta L$, if we were to define the term 'optimize" suitably. In any case these two expressions are closely related, and most statisticians seem to be more comfortable with R.

If A_1 is large the probability of incorrectly rejecting H_1 will be small but at the cost of having the probability of incorrectly accepting H_1 be large. A good strategy involves a compromise between the two error probabilities as well as a good choice of which points properly belong to A_2 rather than A_1, i.e., what observations may be regarded as strong evidence against H_1.

We specialize with the simplest problem in hypothesis testing. We study the test of $H_1 : \theta = \theta_1$ vs. $H_2 : \theta = \theta_2$ from several points of view. It is convenient to note that P_{θ_1} and P_{θ_2} are absolutely continuous with respect to their sum and thus each probability distribution can be represented by a density $f(x|\theta)$ with respect to a common nonnegative finite measure μ. In this case we normally abbreviate $\varepsilon(\theta_i, s) = \varepsilon_i$ and $r(\theta_i) = r_i$.

The Bayesian philosophy attaches a prior probability distribution to the states of nature and P_θ is regarded as the conditional distribution of X given θ. Let $\xi(\theta)$ represent the prior probability of θ. We shall let $1 - \xi(\theta_1) = \xi(\theta_2) = \xi$, $0 \leq \xi \leq 1$. Then the risk for the Bayesian is

$$\mathcal{R}(s) = (1 - \xi)r_1\varepsilon_1 + \xi r_2\varepsilon_2$$

(9.4)
$$= \int_{A_2} (1 - \xi)r_1 f(x|\theta_1)\mu(dx) + \int_{A_1} \xi r_2 f(x|\theta_2)\mu(dx)$$

$$= \int_{A_1} [\xi r_2 f(x|\theta_2) - (1 - \xi)r_1 f(x|\theta_1)]\mu(dx) + (1 - \xi)r_1,$$

which is minimized by putting those points with negative integrand into A_1, i.e.,
  (i) accept H_1 if $\lambda(X) = f(X|\theta_1)/f(X|\theta_2) > \xi r_2/(1 - \xi)r_1 = k$,
  (ii) reject H_1 if $\lambda(X) < k$,
and
  (iii) either if $\lambda(X) = k$.
The statistic

(9.5)
$$\lambda(X) = \frac{f(X|\theta_1)}{f(X|\theta_2)}$$

is called the *likelihood-ratio* and the decision procedure described above is called a *likelihood-ratio test* and treats large values of λ as evidence favoring θ_1. The test is a *Bayes strategy*, which is defined as one that minimizes the *Bayes risk* $\mathcal{R}(s)$ for some prior probability distribution on Θ. Note that k increases with ξ and r_2/r_1. This makes it more difficult to accept H_1 when $\xi = \xi(\theta_2)$ and r_2/r_1 are large. This is intuitively reasonable.

An alternative approach to deriving the Bayes strategy consists of using X to convert the prior distribution ξ to a posterior distribution ξ^* of θ given the data X. Then an action is selected as though ξ^* were the distribution of θ and no data were available. It is relatively trivial to show that this approach is equivalent to the other and this technique, described in more detail below, will be instructive later.

For the case of no data with prior probabilities $1 - \xi$ and ξ, the risks associated with a_1 and a_2 are ξr_2 and $(1 - \xi)r_1$ respectively (see Fig. 5). The Bayes risk is $\min(\xi r_2, (1 - \xi)r_1))$ and is attained *by taking a_1 if $\xi < \xi_0$ and a_2 if $\xi > \xi_0$, where*

(9.6)
$$(1 - \xi_0)r_1 = \xi_0 r_2.$$

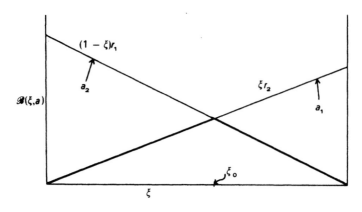

FIG. 5

The posterior probability of θ_2 given X is

$$\xi^* = \frac{\xi f(X|\theta_2)}{(1 - \xi)f(X|\theta_1) + \xi f(X|\theta_2)}$$

(Bayes' theorem) and hence the Bayes strategy consists of taking a_1 if

$$\xi^* = \frac{1}{1 + (1 - \xi)f(X|\theta_1)/\xi f(X|\theta_2)} < \xi_0 = \frac{1}{1 + r_2/r_1},$$

which is equivalent to the rule we derived before. This approach, combined with Bayes' theorem, which states that the *posterior probabilities are proportional to the priors times the likelihoods* and hence

(9.7) $$\frac{1 - \xi^*}{\xi^*} = \frac{1 - \xi}{\xi}\lambda(X),$$

will be useful later.

The Bayesian approach makes the disposal of those x for which $\lambda(x) = k$ relatively unimportant. For non-Bayesians, the set S of all available error points $(\varepsilon_1, \varepsilon_2)$ is of importance. This set contains the points $(0, 1)$ and $(1, 0)$ and by interchanging A_1 and A_2 we have $(1 - \varepsilon_1, 1 - \varepsilon_2) \in S$ if $(\varepsilon_1, \varepsilon_2) \in S$. The Bayes strategies correspond to those points of S which minimize $c_1\varepsilon_1 + c_2\varepsilon_2$ for appropriate nonnegative c_1 and c_2 not both zero (see Fig. 6).

The Lyapunov theorem on the range of a vector measure [B1], [D2] states that S is a closed bounded set, and, if the probability measures have no atoms, S is convex. Thus if X has a continuous distribution under θ_1 and θ_2, S is convex, and the only $(\varepsilon_1, \varepsilon_2)$ which can not be improved are on the *lower boundary* and these correspond to Bayes strategies for appropriate prior probabilities. (This is a special case of a fundamental theorem in decision theory.) If X has a discrete distribution, the set S may consist of a finite set of points and lack convexity.

However, if the statistician uses *randomized strategies*, the line segment connecting any two points that are available is also available. For example, using s_1

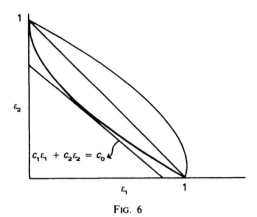

FIG. 6

which yields $(\varepsilon_{11}, \varepsilon_{12})$ with probability $1/3$ and s_2 which yields $(\varepsilon_{21}, \varepsilon_{22})$ with probability $2/3$ is a strategy which yields $(\frac{1}{3}\varepsilon_{11} + \frac{2}{3}\varepsilon_{21}, \frac{1}{3}\varepsilon_{12} + \frac{2}{3}\varepsilon_{22})$. Thus the use of randomized strategies extends S to S^* the convex hull of S. From the class of randomized strategies, the class of Bayes strategies contains those which can not be improved. Thus we have a justification for considering the class of Bayes strategies even if we do not accept the Bayesian philosophy.

There are several alternative ways of randomizing. The method indicated above has the statistician apply a table of random numbers to select one of his *pure* (*nonrandomized*) strategies. Another alternative is to characterize a strategy by a (measurable) function $\varphi(x)$ which indicates the probability of rejecting H_1 if $X = x$ is observed. In this approach a Bayes strategy for testing $H_1 : \theta = \theta_1$ vs. $H_2 : \theta = \theta_2$ is $\varphi(x) = 1$ if $\lambda(x) > k$, $\varphi(x) = 0$ if $\lambda(x) < k$, and $\varphi(x)$ is arbitrary if $\lambda(x) = k$. The type of randomized strategy specified by a measurable function $\varphi(x)$, $0 \leqq \varphi(x) \leqq 1$, is called a *behavioral* strategy.

A third approach is to deatomize potential atoms by observing in addition to X an independent *uniform* random variable U with density $g(u) = 1$ on $0 < u < 1$. Then (X, U) has density $f(x|\theta)$, with no atoms, and $\lambda(X, U) = \lambda(X)$.

One graphically apparent consequence of the use of S^* is that given a prior distribution $(1 - \xi, \xi)$, there is always a nonrandomized Bayes strategy. The Bayes strategies and their mixtures are represented by a point or by a line segment on the lower boundary of S^*.

The original approach to this problem which was formulated by Neyman and Pearson was stated in terms of minimizing ε_2 for given ε_1, a formulation related to the standard convention in statistical practice at that time as well as now, of specifying ε_1, which is called the *significance level* of the test.

A couple of simple examples will terminate this section.

Example 9.1. Cauchy. The random variable X has density

$$f(x|\theta) = \frac{1}{\pi(1 + (x - \theta)^2)},$$

where $H_1 : \theta = \theta_1$ is to be tested versus $H_2 : \theta = \theta_2, \theta_1 > \theta_2$.

The density is symmetric about θ and moving θ simply translates the density. This example is slightly pathological in that it violates a naïve expectation that the likelihood-ratio test would tend to accept H_1 if X is sufficiently large, and to reject otherwise. In fact,

$$\lambda(X) = \frac{1 + (X - \theta_2)^2}{1 + (X - \theta_1)^2} > k$$

defines a set A_1 which is null or a bounded interval if $k > 1$ and two unbounded intervals or the whole space if $k < 1$.

Example 9.2. For $\theta > 0$ let X be a random variable with density

$$f(x|\theta) = \theta e^{-\theta x}, \quad x > 0,$$

$$= 0, \qquad x < 0.$$

To test $H_1 : \theta = \theta_1$ vs. $H_2 : \theta = \theta_2, \theta_1 < \theta_2$, we observe a sample of n independent observations $X_1, X_2, \cdots, X_n$ on X (i.e., $X_1, X_2, \cdots, X_n$ are i.i.d. with the same distribution as X). Let $Z = (X_1, X_2, \cdots, X_n)$. Then

$$f(z|\theta) = \prod_{i=1}^{n} (\theta e^{-\theta x_i})$$

and

$$\log \lambda(Z) = n[\log(\theta_1/\theta_2) + (\theta_2 - \theta_1)\bar{X}]$$

and the likelihood-ratio tests consist of accepting H_1 if $\lambda(Z) > k$ or if

$$\bar{X} > k'$$

for some appropriate constant k'.

That a large value of the sample mean $\bar{X}$ should lead to the acceptance of H_1 is reasonable in view of the fact that $E(X) = \theta^{-1}$ decreases with θ. The fact that the form of this test does not depend on the particular values of θ_1 and θ_2 as long as $\theta_1 < \theta_2$ implies that tests of this form are very good for testing $H_1^* : \theta \leq \theta_0$ vs. $H_2^* : \theta > \theta_0$.

Note that as $n \to \infty$, the ability to discriminate between θ_1 and θ_2 increases. Thus the set of available $(\varepsilon_1, \varepsilon_2)$ comes closer and closer to the origin. In fact using $k' = (\theta_1^{-1} + \theta_2^{-1})/2$, both ε_1 and ε_2 approach zero.

9.1. Large sample considerations. We now introduce a variation of Example 9.2 which has some experimental design aspects.

Example 9.1.1. Suppose that X represents the lifetime of a light bulb under a high stress environment. It is assumed to have an exponential distribution with *failure rate* θ, i.e., the density is $\theta \exp(-\theta x)$. Rather than observe the test equipment for n bulbs continuously until all the bulbs fail, two times of observation t_1 and t_2 are set up. By checking at these times, we can count the number of bulbs that failed between 0 and t_1, between t_1 and t_2 and how many will fail after t_2.

As an estimation problem this is amenable to study by means of the Fisher information. If time is valuable one could measure the amount of information divided by the cost of the experiment which would be affected by t_2. For the moment, let us assume that time is not a factor, that we are interested in testing $H_1 : \theta = \theta_1$ vs. $H_2 : \theta = \theta_2$, $\theta_1 < \theta_2$ and that practical considerations force us to base our test on the statistic

$$T = n_1 + n_2,$$

where n_i is the number of bulbs still alive at time t_i, $i = 1\ 2$. It is more convenient to write

$$T = m_1 + 2m_2,$$

where m_1 is the number of bulbs that fail between t_1 and t_2 and $m_2 = n_2$. This convenience derives from the fact that it is now clear that T is a sum of *independent* Y_i, i.e.,

$$T = \sum_{i=1}^{n} Y_i = n\overline{Y},$$

where $Y_i = 0$, 1 or 2 depending on which of the time intervals, $(0, t_1)$, (t_1, t_2), (t_2, ∞) is the one in which X_i falls.

It is not unusual to find many problems where convenience leads to the use of test statistics which are sums of i.i.d. random variables. The likelihood-ratio test based on n independent observations is such a test for

(9.1.1) $$\log \lambda(Z) = \sum_{i=1}^{n} \log[f(X_i|\theta_1)/f(X_i|\theta_2)] = \sum_{i=1}^{n} Y_i = n\,\overline{Y},$$

where $Z = (X_1, X_2, \cdots, X_n)$ is a sequence of n i.i.d. variables, and Y_i is the logarithm of the likelihood-ratio based on X_i.

To evaluate a test procedure based on such a *sum* or *average* requires the distribution of the sum or average. The *central limit theorem* tells us that $\overline{Y}$ is approximately normally distributed with mean μ_Y and variance σ_Y^2/n. More precisely,

$$\mathcal{L}[\sqrt{n}(\overline{Y} - \mu_Y)/\sigma_Y] \to \mathcal{N}(0, 1).$$

Thus, the use of the procedure "accept H_1 if $\overline{Y} \geq k$" has probability of rejection

$$\alpha(\theta_i) \approx \Phi\left[\sqrt{n}\frac{-[k - \mu_Y(\theta_i)]}{\sigma_Y(\theta_i)}\right], \qquad i = 1, 2.$$

However, if $\mu_Y(\theta_1) > k > \mu_Y(\theta_2)$, as $n \to \infty$, this simply states that $\alpha(\theta_1) \to 0$ and $\alpha(\theta_2) \to 1$ which is desirable and expected but rather crude. It would be desirable to have a more refined estimate of how rapidly $\alpha(\theta_i) \to 0$ or 1. Unfortunately the central limit theorem does not specify how rapidly the probability approaches zero in the *tails* of the distribution.

To cope with this question, an extension of the central limit theorem is required. Interestingly enough, a rather coarse approximation is quite adequate for some basic results.

THEOREM 9.1.1. *If* $X_1, X_2, \cdots, X_n$ *are i.i.d. with mean* $\mu_X > k$, *then*

(9.1.2) $$\frac{1}{n} \log P(\overline{X} \leq k) \to \inf_t \log E(e^{t(X-k)}).$$

This theorem states that if $k < \mu_X$, $P(\overline{X} \leq k)$ approaches 0 roughly exponentially in n, i.e., as $\exp(-n\rho)$, where $\rho = -\log \inf_t E \exp[t(X - k)]$. The term "rough" is appropriate since multiplying $P(\overline{X} \leq k)$ by a power of n would not affect the left-hand side of (9.1.2). By replacing X by $-X$, we see that for $k > \mu_X$,

$$\frac{1}{n} \log P(\overline{X} \geq k) \to \inf_t \log E(e^{t(X-k)}).$$

We shall start with a simple lemma of the Chebyshev type. But first a few words about moment generating functions which are essentially Laplace transforms. The *moment generating function* (m.g.f.) of a random variable X is defined by

$$M_X(t) = E(e^{tX}).$$

The name comes from the fact that

$$M_X(t) = 1 + tE(X) + \cdots + \frac{t^r E(X^r)}{r!} + \cdots$$

when the expansion is valid. We have $M_X(0) = 1$. If $M_X(a) < \infty$ and $M_X(b) < \infty$ for real a and $b, a < b$, then $M_X(t)$ is finite for all complex t in the slab $a \leq \mathrm{Re}(t) \leq b$ and $M_X(t)$ is analytic in the interior of this slab. $M_X(t)$ is convex in $[a, b]$ and $d^r M_X(t)/dt^r = E(X^r e^{tX})$ exists finite for $a < t < b$. $M_X(t)$ is strictly convex unless $X = 0$ w.p.1. If $\mu = E(X) > 0$, then $M_X(t) > 1$ for $t > 0$. If X and Y are independent, then

$$M_{X+Y}(t) = M_X(t)M_Y(t).$$

It is easy to see that for i.i.d. X_i,

$$M_{\overline{X}}(t) = \left[M_X\!\left(\frac{t}{n}\right) \right]^n$$

and

$$M_{X-k}(t) = e^{-tk}M_X(t).$$

LEMMA 9.1.1. *If* $X_1, X_2, \cdots, X_n$ *are i.i.d. observations on a r.v.* X *and* $k \leq E(X)$, *then*

(9.1.3) $$P(\overline{X} \leq k) \leq [M_{X-k}(t)]^n \quad \text{for all} \quad t.$$

Proof. First we show that if $E(Y) \geq 0$,

(9.1.4) $$P(Y \leq 0) \leq M_Y(t) \quad \text{for all} \quad t.$$

For $t \geq 0$, $M_Y(t) \geq 1$ and this result is trivial. For $t \leq 0$, $\exp(tY) \geq 1$ for $Y \leq 0$ and $\exp(tY) > 0$ for $Y > 0$. Hence $\exp(tY) \geq I_{(Y \leq 0)}$, where $I_A = 1$ on A and 0

elsewhere. Taking expectations yields (9.1.4) immediately. Now let $Y = \overline{X} - k$, and observe that $E(Y) = E(X) - k \geq 0$ and

$$M_Y(t) = \left[M_{X-k}\left(\frac{t}{n}\right) \right]^n.$$

The lemma follows.

Lemma 9.1.1 provides an upper bound on $P(\overline{X} \leq k)$ which gives half of the theorem. Incidentally since $E \exp[t(X - k)] > 1$ for $t > 0$, the $\inf_t$ in the statement can be confined to $\inf_{t \leq 0}$.

A proof of the other half appears in [C5]. However, in the present form the theorem is rather crude though adequate for our needs. More refined results were obtained by Cramér [C6]. Two techniques are quite useful for these more refined results which yield $P(\overline{X} \leq k) \approx Kn^{-1/2} e^{-n\rho}$ under suitable regularity conditions. One technique is the method of steepest descent applied to the inversion of the characteristic function of $\overline{X}$

$$\varphi_{\overline{X}}(t) = E(e^{it\overline{X}}).$$

(The characteristic function of a random variable is analogous to the Fourier transform and can be inverted to determine the distribution of the random variable.) A second method used by Cramér consists of *tilting* the distribution by means of the transformation:

$$dF^*(x) = e^{tx} \, dF(x)/M_X(t)$$

so that the tilted distribution has mean k. Then there is a similar relationship between the distribution functions of $\overline{X}$ before and after tilting and the central limit theorem applied to F^* yields refined results for the tail of $\overline{X}$ under F. The tilting method requires $M_X(t)$ to exist away from $t = 0$ and is affected if F is a lattice distribution.

Both of these techniques are useful for finding asymptotic expansions for $P(\sqrt{n}(\overline{X} - \mu_X) \leq k_n)$, where $k_n \to \infty$ no faster than $\sqrt{n}$.

If $M'_X(t)$ exists at the minimizing value t_0 of $M_{X-k}(t)$ then, by differentiation,

(9.1.5)
$$M'_{X-k}(t_0) = E[(X - k) e^{t_0(X-k)}] = 0,$$
$$k = E(Xe^{t_0 X})/M_X(t_0),$$

which is easily solved numerically for t_0 since the right-hand side is monotone in t_0.

This equation may be applied to Example 9.1.1. If $Y = 0$, 1 and 2 with probabilities p_0, p_1 and p_2,

$$k = (p_1 e^{t_0} + 2p_2 e^{2t_0})/(p_0 + p_1 e^{t_0} + p_2 e^{2t_0}).$$

To be specific, suppose that $\theta_1 = 1$, $\theta_2 = 1.2$. Then for each t_1 and t_2, Y has an expectation $\mu_Y(\theta; t_1, t_2)$ with $\mu_Y(\theta_1; t_1, t_2) > \mu_Y(\theta_2; t_1, t_2)$. Let k be an arbitrary value between these. Confining attention to "one-tailed" tests based on Y, we find, as $n \to \infty$,

(9.1.6) $$\varepsilon_1 = \varepsilon(\theta_1; t_1, t_2, k) = P_{\theta_1}(\bar{Y} \le k) \sim e^{-n\psi_1}$$

and

(9.1.7) $$\varepsilon_2 = \varepsilon(\theta_2; t_1, t_2, k) = P_{\theta_2}(\bar{Y} > k) \sim e^{-n\psi_2},$$

where $\sim$ indicates the rough approximation in the sense that $-n^{-1}\log \varepsilon_i \to \psi_i$, and ψ_1 decreases in k and ψ_2 increases in k.

A Bayesian might be interested in minimizing $\zeta_1 r_1 \varepsilon_1 + \zeta_2 r_2 \varepsilon_2$. Clearly he would select k so as to maximize $[\min(\psi_1, \psi_2)]$. The best experimental choice corresponds to that value of (t_1, t_2) for which the above maximum is maximized.

Of course the discussion of this example is applicable to a more general situation. Essentially our asymptotic results, which are commonly called *large-deviation* results because they correspond to a large number of standard deviations from the mean of $\bar{Y}$, provide a measure of how good an experiment is for deciding between $H_1 : \theta = \theta_1$ and $H_2 : \theta = \theta_2$ by using one-sided tests based on the average $\bar{Y}$ of i.i.d. observations on a random variable Y. This measure is

(9.1.8) $$I = \max_{\mu_2 \le k \le \mu_1} [\min[\psi_1(k), \psi_2(k)]],$$

where $\mu_i = E_{\theta_i}(Y)$, $\mu_1 > \mu_2$ and

(9.1.9) $$\psi_i(k) = -\log\{\inf_t E_{\theta_i}[e^{t(Y-k)}]\}.$$

For the appropriate choice of k, we have a test for which $\varepsilon(\theta_i) \sim \exp(-nI)$, $i = 1, 2$. This choice of I was justified above on Bayesian grounds. It is of interest to note that I does not depend on the particular values of $\zeta_1, \zeta_2, r_1, r_2$ as long as they are positive. In other words, for large samples, the precise choice of losses and prior probabilities has relatively little effect on the optimal procedures.

For a non-Bayesian justification, we may consider the "objective" statistician who wishes to minimize some objective criterion $G(\varepsilon_1, \varepsilon_2)$, where G is locally linear for $(\varepsilon_1, \varepsilon_2)$ near $(0, 0)$, i.e., $\partial G/\partial \varepsilon_1 > 0$ and $\partial G/\partial \varepsilon_2 > 0$. Here again I is relevant. If, however, we wish to minimize

$$G(\varepsilon_1, \varepsilon_2) = c_1 \varepsilon_1^2 + c_2 \varepsilon_2,$$

which might be appropriate in cases where the importance of the two different types of error are of different orders of magnitude, a different measure becomes appropriate, i.e., $\max_k[\min(2\psi_1, \psi_2)]$.

Before leaving this problem of testing $H_1 : \theta = \theta_1$ vs. $H_2 : \theta = \theta_2$, there are two additional questions of interest. First the likelihood-ratio tests are based on averages of random variables. How does our measure of information I behave in that case? Second, in some special applications and in classical theory much attention is paid to the case when ε_1 is fixed and $\varepsilon_2 \to 0$. Does the large deviation theory give rise to another appropriate information number here?

The likelihood-ratio tests consist of rejecting $H_1 : \theta = \theta_1$ if

$$\bar{Y} \leqq k,$$

where $\bar{Y}$ is the average of the observations on

$$(9.1.10) \qquad Y = \log[f_1(X)/f_2(X)]$$

(using $f_i(\dot{X})$ for $f(X|\theta_i)$). Under the hypothesis H_1, Y has mean

$$(9.1.11) \qquad \mu_Y(\theta_1) = E_{\theta_1} \log[f_1(X)/f_2(X)].$$

We establish the following lemma.

LEMMA 9.1.2. $\mu_Y(\theta_1) \geqq 0$, with strict inequality unless $P_{\theta_1} = P_{\theta_2}$.
Proof. Since $-\log x$ is a strictly convex function,

$$E_{\theta_1}(-\log Z) \geqq -\log E_{\theta_1}(Z)$$

as long as $E_{\theta_1}(Z)$ exists (possibly infinite), with strict inequality unless Z is constant with probability one. If $Z = f_2(X)/f_1(X)$, the left-hand side is $\mu_Y(\theta_1)$ and the right-hand side is $-\log P_{\theta_2}(f_1(X) > 0) \geqq 0$.

To achieve $\mu_Y(\theta_1) = 0$, it is necessary that Z be constant with P_{θ_1} probability 1 and $P_{\theta_2}(f_1(X) > 0) = 1$. These conditions imply that $\mu_Y(\theta_1) > 0$ unless $P_{\theta_1} = P_{\theta_2}$. Thus unless $P_{\theta_1} = P_{\theta_2}$, an exceptional case in which nothing useful can happen and which we exclude, $\mu_Y(\theta_1) > 0$.

Reversing f_1 and f_2, this lemma yields the inequality $\mu_Y(\theta_2) \leqq 0$ with strict inequality unless $P_{\theta_1} = P_{\theta_2}$.

THEOREM 9.1.2. Using $Y = \log[f_1(X)/f_2(X)]$, the measure of information I is given by

$$(9.1.12) \qquad I = -\log \inf_{0 \leqq t \leqq 1} \int f_1^t(x) f_2^{1-t}(x)\mu(dx).$$

Proof. For simplicity let us assume that $f_1(x)$ and $f_2(x)$ are both positive on the same set and $P_{\theta_1} \neq P_{\theta_2}$. Since $\psi_1(k)$ and $\psi_2(k)$ are monotonic in k it will suffice to show that they coincide at the above value for $k = 0$. But

$$(9.1.13) \qquad \psi_1(0) = -\log \inf_t E_{\theta_1}(e^{tY}) = -\log \inf_t \int f_1^{1+t}(x) f_2^{-t}(x)\mu(dx).$$

Since $\mu_Y(\theta_1) > 0$, the infimum is obtained for $t \leqq 0$. Also

$$(9.1.13') \qquad \psi_2(0) = -\log \inf_t E_{\theta_2}(e^{tY}) = -\log \inf_t \int f_1^t(x) f_2^{1-t}(x)\mu(dx),$$

where this infimum is attained for $t \geqq 0$ since $\mu_Y(\theta_2) < 0$. Transforming t to $t' - 1$ in (9.1.13), we have $\psi_1(0) = \psi_2(0) = I$ and the desired result.

The measure I may be regarded as a type of distance between θ_1 and θ_2 or as a measure of *information*. One reason to call it an information is that if an experiment with outcome X^* yields $I^* = 2I$, then comparable ε_1 and ε_2 are attainable with half as many observations, i.e., $\exp(-I^* n^*) = \exp(-In)$ for $n^* = n/2$. (The Fisher information had a comparable property with respect to its relation to variance or mean squared error.) It follows, and can easily be checked directly, that we may write $I_{(X,X)} = 2I_X$. But unlike the Fisher information, if Y and Z are independent,

$$I_{(Y,Z)} \leqq I_Y + I_Z.$$

We illustrate the measure I with two examples.

Example 9.1.2. Binomial distribution.

$$P(X = x) = p^x (1 - p)^{1-x}, \qquad\qquad x = 0, 1.$$

For testing $H_1 : p = p_1$ vs. $H_2 : p = p_2$, where $p_1 > p_2$, it is easy to see that the likelihood-ratio statistic is monotone in $\hat{p}$ which is the proportion of observations where $X = 1$. A complicated computation yields

(9.1.14)
$$I = c \log \frac{c}{p_1} + (1 - c) \log \left(\frac{1 - c}{1 - p_1} \right),$$

where

$$c = \frac{\log[(1 - p_1)/(1 - p_2)]}{\log[(1 - p_1)/(1 - p_2)] + \log[p_2/p_1]}.$$

Example 9.1.3. Normal distributions with known variances. Let X have the normal distribution $\mathcal{N}(\mu, \sigma^2)$, where under $H_i : \mu = \mu_i$, $\sigma^2 = \sigma_i^2$, $i = 1, 2$ and $\mu_1 > \mu_2$.

The one-sided test based on $\bar{X}$ will *not* be a likelihood-ratio test if $\sigma_1^2 \neq \sigma_2^2$. However,

$$M_{X-k}(t) = E_{\theta_i}(e^{t(X-k)}) = e^{t(\mu_i - k) + t^2 \sigma_i^2 / 2},$$

$$\inf_t M_{X-k}(t) = \exp[-(\mu_i - k)^2 / 2\sigma_i^2].$$

It follows readily that using $\bar{X}$ instead of the likelihood-ratio test,

(9.1.15)
$$I = \frac{1}{2} \frac{(\mu_1 - \mu_2)^2}{(\sigma_1 + \sigma_2)^2}.$$

This could have been easily derived by noting that the test which rejects H_1 if $\bar{X} \leqq k$ has $\varepsilon_1 = \Phi[\sqrt{n}(k - \mu_1)/\sigma_1]$ and $\varepsilon_2 = \Phi[\sqrt{n}(\mu_2 - k)/\sigma_2]$ and these error probabilities match at $\Phi[-\sqrt{n}(\mu_1 - \mu_2)/(\sigma_1 + \sigma_2)]$ when $k = (\mu_1 \sigma_1^{-1} + \mu_2 \sigma_2^{-1})/(\sigma_1^{-1} + \sigma_2^{-1})$.

9.2. Kullback–Leibler information numbers. Let us consider the special problem of letting $\varepsilon_2 \to 0$ as ε_1 is kept fixed at some nominal level (say 0.05). While this is a classical form often used in hypothesis testing its relevance in *large sample theory* is relatively rare and confined to situations where one type of error is tolerable but the frequency or probability of another must be kept minimal.

The large deviation result provides a mechanism for dealing with this problem. Let us confine attention to the likelihood-ratio test. Then, for a test of the form "reject $H_1 : \theta = \theta_1$ if $\bar{Y} \leq k_n$" to have ε_1 bounded away from 0 and 1, we need $k_n \to E_{\theta_1}(Y) = \mu_Y(\theta_1)$. Then $\varepsilon_2 \sim \exp[-n\psi_2(k)]$, where $k = \mu_Y(\theta_1)$. However, by checking the derivative with respect to t, we see that

$$E_{\theta_2}[e^{t[Y - \mu_Y(\theta_1)]}]$$

achieves a minimum at $t = 1$ and this minimum is $\exp(-I_{12})$, where

(9.2.1) $$I_{12} = I(f_1, f_2) = \mu_Y(\theta_1) = \int \log[f_1(x)/f_2(x)] f_1(x) \mu(dx)$$

is called the *Kullback–Leibler information number* for discriminating between f_1 and f_2 when f_1 represents the true distribution. The preceding Lemma 9.1.2 indicated that $I(f_1, f_2) \geq 0$ with equality only if $f_1 = f_2$ a.e. $[\mu]$. As we have indicated above, $\varepsilon_2 \sim \exp[-nI(f_1, f_2)]$ if ε_1 is bounded away from 0 and 1. Similarly if ε_2 is bounded away from 0 and 1, we have $\varepsilon_1 \sim \exp[-nI(f_2, f_1)]$, where $I(f_2, f_1) = -\mu_Y(\theta_2)$ is ordinarily *not* equal to $I(f_1, f_2)$.

In those cases where f_2 can vanish on a set of positive P_{θ_1} probability, $I(f_1, f_2)$ becomes infinite. It is clear from the interpretation in terms of error probabilities of the three measures $I(f_1, f_2)$, $I(f_2, f_1)$ and $I = -\log \inf_{0 \leq s \leq 1} \int f_1^s(x) f_2^{1-s}(x) \mu(dx)$, that each of the two Kullback–Leibler information numbers $\geq I$. On the other hand, a change of experiment which increases one of these three numbers need not increase the others. All three information numbers may be regarded as functions of the $(\varepsilon_1, \varepsilon_2)$ curves based on the likelihood-ratio tests using a single observation X.

Returning to the examples of the preceding section we see that for
Example 9.1.2 (binomial),

(9.2.2) $$I(p_1, p_2) = p_1 \log \frac{p_1}{p_2} + (1 - p_1) \log \left[\frac{1 - p_1}{1 - p_2}\right],$$

and for
Example 9.1.3 (normal),

(9.2.3) $$I(\theta_1, \theta_2) = \frac{1}{2}\left[\frac{(\mu_1 - \mu_2)^2}{\sigma_2^2} + \frac{\sigma_1^2}{\sigma_2^2} - \log \frac{\sigma_1^2}{\sigma_2^2} - 1\right].$$

Note that $I(p, 0) = I(p, 1) = \infty$ for $0 < p < 1$ but $I(0, p) = -\log(1 - p)$ and $I(1, p) = -\log p$. This is explained by letting p be the probability of success of an event. If $p \neq 0$, eventually a success occurs which *proves* $p \neq 0$. Hence $I(p, 0) = \infty$. On the other hand if $p = 0$, n successive failures is strong but not definitive evidence that p is not positive. Hence $I(0, p) \neq \infty$. The term $\sigma_1^2/\sigma_2^2 - \log(\sigma_1^2/\sigma_2^2) - 1$ which adjusts $I(\theta_1, \theta_2)$ to compensate for $\sigma_1 \neq \sigma_2$ is always positive unless $\sigma_1 = \sigma_2$.

The Kullback–Leibler information number is additive in the sense that

(9.2.4) $$I_{X,Y}(\theta_1, \theta_2) = I_X(\theta_1, \theta_2) + I_Y(\theta_1, \theta_2),$$

where $I_X(\theta_1, \theta_2)$ represents the information using the distributions of X when $\theta = \theta_1$ and $\theta = \theta_2$ respectively and X and Y are independent. Moreover, $I(\theta_1, \theta_2)$

is linear under randomization. The experiment which uses X with probability λ and Y with probability $(1 - \lambda)$ has information $\lambda I_X(\theta_1, \theta_2) + (1 - \lambda) I_Y(\theta_1, \theta_2)$.

9.3. A multivariate normal example. In recent years there has been a considerable interest in dealing with multivariate data of high dimensionality. Problems in classification and pattern recognition involve experiments with so much data per observation that one seeks methods of selecting a few of these variables or functions of these variables which carry much of the relevant information. When one selects a few of the available observations, that choice effectively constitutes a choice of experiment.

While the multivariate normal distribution has dubious value in high dimensional problems, it represents a distribution in which the analysis is reasonably simple and whose results may be suggestive. Let us evaluate I and $I(\theta_1, \theta_2)$ for the multivariate normal case. Here X has the $\mathcal{N}(\mu, \Sigma)$ distribution where under hypothesis $H_i : \theta = (\mu, \Sigma) = (\mu_i, \Sigma_i)$, $i = 1, 2$, where μ_i is a $(k \times 1)$ vector and Σ_i is a $(k \times k)$ positive definite symmetric matrix.

We recall that the normal density is

$$(9.3.1) \qquad (2\pi)^{-k/2} |\Sigma|^{-1/2} \exp[-(x - \mu)'\Sigma^{-1}(x - \mu)/2].$$

A simple generalization of (9.2.3) yields

$$(9.3.2) \quad I(\theta_1, \theta_2) = \frac{1}{2}\left[(\mu_1 - \mu_2)'\Sigma_2^{-1}(\mu_1 - \mu_2) + \log\frac{|\Sigma_2|}{|\Sigma_1|} - k + \operatorname{tr} \Sigma_2^{-1}\Sigma_1 \right].$$

Before deriving I, let us extend the information of Example 9.1.3. In § 9.1 we first confined attention to one-sided tests based on $\overline{Y}$ rather than on likelihood-ratio tests. An analogue for k-dimensional random variables X, is to study one-sided tests based on linear functions of $\overline{X}$, or equivalently on averages of linear functions of X. (The term *threshold* is applied to such tests in the electrical engineering literature.) However, under H_i, the linear function $Y = a'X$ has a univariate normal distribution with mean $v_i = a'\mu_i$ and variance $\tau_i^2 = a'\Sigma_i a$. Thus the rate at which ε_1 and ε_2 approach zero when testing is based on one-sided tests using linear functions of $\overline{X}$ is determined by

$$(9.3.3) \qquad I = \frac{1}{2}\sup_a \left(\frac{v_1 - v_2}{\tau_1 + \tau_2}\right)^2 = \frac{1}{2}S^2,$$

where

$$S = \sup_a \frac{|a'(\mu_1 - \mu_2)|}{(a'\Sigma_1 a)^{1/2} + (a'\Sigma_2 a)^{1/2}}.$$

This measure involves a maximization with respect to k variables which looks potentially difficult. However, it can also be expressed as a maximum with respect to a single variable.

Applying the method of Lagrange multipliers to minimize $(a'\Sigma_1 a)^{1/2} + (a'\Sigma_2 a)^{1/2}$ subject to $a'(\mu_1 - \mu_2) = c$, we have

$$\frac{\Sigma_1 a}{(a'\Sigma_1 a)^{1/2}} + \frac{\Sigma_2 a}{(a'\Sigma_2 a)^{1/2}} = \rho(\mu_1 - \mu_2).$$

Let

$$\Sigma = t\Sigma_1 + (1-t)\Sigma_2,$$

where t and $(1-t)$ are proportional to $(a'\Sigma_1 a)^{-1/2}$ and $(a'\Sigma_2 a)^{-1/2}$. Since S is unaffected when a is multiplied by a constant, we may take $a = \Sigma^{-1}(\mu_1 - \mu_2)$. The following theorem due to Anderson and Bahadur [A8] and Clunies-Ross and Riffenburgh [C19] follows.

THEOREM 9.3.1.

(9.3.4) $$S^2 = t(1-t)[(\mu_1 - \mu_2)'\Sigma^{-1}(\mu_1 - \mu_2)],$$

where

(9.3.5) $$\Sigma = t\Sigma_1 + (1-t)\Sigma_2$$

and t is the unique root between 0 and 1 of

(9.3.6) $$R(t) = (\mu_1 - \mu_2)'\Sigma^{-1}[t^2\Sigma_1 - (1-t)^2\Sigma_2]\Sigma^{-1}(\mu_1 - \mu_2) = 0.$$

Furthermore,

(9.3.7) $$R'(t) = 2(\mu_1 - \mu_2)'\Sigma^{-1}\Sigma_1\Sigma^{-1}\Sigma_2\Sigma^{-1}(\mu_1 - \mu_2) > 0$$

and a maximizing choice of a is given by

(9.3.8) $$a = \Sigma^{-1}(\mu_1 - \mu_2).$$

Finally let us consider the information I based on the use of the likelihood-ratio. Then, a moderately complex calculation yields the next theorem [C18].

THEOREM 9.3.2.

$$I = \tfrac{1}{2}T^2,$$

where

(9.3.9) $$T^2 = \sup_{0 \le t \le 1} \{t(1-t)(\mu_1 - \mu_2)'\Sigma^{-1}(\mu_1 - \mu_2) + \log|\Sigma|$$

$$- [t\log|\Sigma_1| + (1-t)\log|\Sigma_2|]\},$$

(9.3.10) $$\Sigma = t\Sigma_1 + (1-t)\Sigma_2,$$

and if the expression in braces in (9.3.9) is $H(t)$,

$$H'(t) = (\mu_1 - \mu_2)'\Sigma^{-1}[(1-t)^2\Sigma_2 - t^2\Sigma_1]\Sigma^{-1}(\mu_1 - \mu_2)$$

$$- \log|\Sigma_1\Sigma_2^{-1}| + \operatorname{tr}(\Sigma_1 - \Sigma_2)\Sigma^{-1},$$

and

$$H''(t) = -2(\mu_1 - \mu_2)'\Sigma^{-1}\Sigma_1\Sigma^{-1}\Sigma_2\Sigma^{-1}(\mu_1 - \mu_2)$$

$$- \operatorname{tr}(\Sigma_1 - \Sigma_2)\Sigma^{-1}(\Sigma_1 - \Sigma_2)\Sigma^{-1} < 0.$$

The concavity of log determinant implies that $T^2 \geq S^2$ which is anticipated from the fact that T^2 is based on the likelihood-ratio rather than the best *linear discriminant* function.

An illustration which may be regarded as mildly informative concerns the following situation. Suppose that $X_1, X_2, \cdots, X_{2n}$ are i.i.d. normal with mean μ_i and variance σ_i^2 under H_i, $i = 1, 2$, where $\mu_1 = 1$, $\mu_2 = -1$, $\sigma_1 = 1$ and $\sigma_2 = 3$. Let $Y_1, Y_2, \cdots, Y_n$ be i.i.d. with mean μ_i and variance τ_i^2 under H_i, $i = 1, 2$, where $\tau_1 = 3, \tau_2 = 1$. Suppose that it is decided to use X_i, $1 \leq i \leq n$, and a choice must be made between X_i, $n + 1 \leq i \leq 2n$, on the one hand and Y_i, $1 \leq i \leq n$, on the other.

Informally the X's are more precise under H_1 than under H_2 and the reverse is true for the Y's. In this testing context which goes under the names of discrimination and classification, should we prefer to balance the first nX's with Y's or to maintain the imbalance? For both the linear and nonlinear cases, there is some advantage in maintaining a lack of balance and concentrating on experiments in which the additional data are more precise under H_1. (See Table 1.)

TABLE 1

	$2nX$'s	nX', nY's
$S^2/2n$	0.25	0.20
$T^2/2n$	0.81	0.71

The normal distribution example where $\Sigma_1 = \Sigma_2 = \Sigma$ is a particularly simple and useful case. Then the likelihood-ratio is linear and the corresponding discriminant function

$$(9.3.11) \qquad Y = (\mu_1 - \mu_2)'\Sigma^{-1}X$$

or for symmetry one can use $(\mu_1 - \mu_2)'\Sigma^{-1}(X - (\mu_1 + \mu_2)/2)$. The measure D, given by

$$(9.3.12) \qquad D^2 = (\mu_1 - \mu_2)'\Sigma^{-1}(\mu_1 - \mu_2) = 4S^2 = 4T^2 = 8I$$

is called the *Mahalanobis distance*. One way to remember the interpretation of D is in terms of the example where $\Sigma = I$, μ_1 and μ_2 differ by δ in one coordinate and are equal in all others. Then $D = \delta$.

9.4. Consistency of the maximum likelihood estimate. In the history of the m.l.e., the asymptotic normality property given consistency was relatively easy to get straight. The preliminary result of consistency gave more difficulty. We shall present here a proof based on one by Wald [W1].

This section is somewhat of a digression from the main theme of these notes and was delayed for two reasons. The use of the Kullback–Leibler information suggested delaying it till we discussed that. At the early point where m.l.e. was brought up it would have seemed another technicality piled on several others with a tendency to divert attention from the main business.

Proof of consistency. Let $X_1, X_2, \cdots$ be independent observations on X with density $f(x|\theta)$, where $\theta \in \Theta$. To prove consistency of the maximum-likelihood estimate $\hat{\theta}_n$ of θ, we consider first the special case where Θ is finite and then, using compactness, extend the proof to derive a result for infinite sets Θ.

Suppose Θ is finite and $\theta_0 \in \Theta$ is the true state of nature, i.e., all probabilities and expectations will be with respect to θ_0 unless otherwise specified. Then the m.l.e. $\hat{\theta}_n$ is equal to θ_0 unless

$$\prod_{i=1}^{n} f(X_i|\theta_0) \leq \prod_{i=1}^{n} f(X_i|\varphi) \quad \text{for some} \quad \varphi \in \Theta.$$

Then

$$\sum_{i=1}^{n} Y_i \leq 0,$$

where

$$Y_i = \log[f(X_i|\theta_0)/f(X_i|\varphi)]$$

has positive mean $E_{\theta_0}(Y_i) = I(\theta_0, \varphi) > 0$ as long as $P_\varphi \neq P_{\theta_0}$. But the law of large numbers states that $\bar{Y}_n = n^{-1} \sum_{i=1}^{n} Y_i$ converges to $I(\theta_0, \varphi)$ with probability one. This implies that with probability one the likelihood of θ_0 eventually exceeds that of φ for each $\varphi \neq \theta_0$. Thus with probability one, $\hat{\theta}_n = \theta_0$ for $n \geq N$ (where N may be random).

Thus we have proved a *strong consistency* for the case where $P_\varphi \neq P_{\theta_0}$ if $\varphi \neq \theta_0$. This implies the weaker statement that $\hat{\theta}_n \to \theta_0$ in probability as $n \to \infty$. Indeed $P_{\theta_0}(\hat{\theta}_n \neq \theta_0) \to 0$ as $n \to \infty$.

For the case where Θ is infinite, we shall try to establish that $P\{|\hat{\theta}_n - \theta_0| > \varepsilon\} \to 0$ as $n \to \infty$ for each $\varepsilon > 0$. To do so we shall cover the complement of the ε-neighborhood of θ_0 by a finite number of relatively small intervals in each of which P_φ is almost constant. To do this requires a compactness condition which is not satisfied in as simple an example as $\mathscr{L}(X) = \mathscr{N}(\mu, 1)$, $-\infty < \mu < \infty$. On the other hand,

$$f(x|\mu) = (2\pi)^{-1/2} \exp[-(x - \mu)^2/2] \to 0 \quad \text{as} \quad \mu \to \infty$$

and the space $-\infty < \mu < \infty$ may be compactified by adjoining a point at ∞ to Θ for which the density is identically 0. However, this is not the density of a random variable for its integral is less than one. But it corresponds to a *subdistribution* with total measure less than one.

Thus we shall impose the condition on Θ that it can be compactified by adjoining additional elements to obtain Θ^* so that for each $\varphi \in \Theta^*$ there is a *subdistribution* P_φ with density $f(x|\varphi)$ such that $f(x|\varphi)$ is continuous in φ and $\int f(x|\varphi)\mu(dx) \leq 1$. Of course, to talk of compactness and continuity we need a topology on Θ^*.

Lemma 9.1.2 states that $I(\theta_1, \theta_2) > 0$ unless $P_{\theta_1} = P_{\theta_2}$ when P_{θ_1} and P_{θ_2} are distributions. It is easy to see that if P_φ is a subdistribution,

$$(9.4.1) \qquad I(\theta_0, \varphi) = \int f(x|\theta_0) \log[f(x|\theta_0)/f(x|\varphi)]\mu(dx) > 0 \quad \text{unless} \quad P_\varphi = P_{\theta_0}.$$

We are ready to prove the following theorem.

THEOREM 9.4.1. *Consistency of maximum likelihood estimates. If it is possible to extend* Θ *to a compact set* Θ^* *such that*

(i) *for every* $\varphi \in \Theta^*$, P_φ *is a subdistribution with density* $f(x|\varphi)$ *which is continuous with respect to* φ,

(ii) *for every* $\varphi \in \Theta^*$, *there is a neighborhood* N_φ *of* φ *such that*

$$E_{\theta_0}[\inf_{\varphi' \in N_\varphi} \log[f(X|\theta_0)/f(X|\varphi')]] > -\infty,$$

and

(iii) *for every* $\varphi \in \Theta^* - \{\theta_0\}$, $P_\varphi \neq P_{\theta_0}$,

then, for every neighborhood N_{θ_0} *of* θ_0, $\hat{\theta}_n$, *the value of* $\varphi \in \Theta^*$ *which maximizes the likelihood based on the first n observations, exists and satisfies*

$$P_{\theta_0}(\hat{\theta}_n \notin N_{\theta_0} \text{ infinitely often}) = 0$$

and

$$P_{\theta_0}(\hat{\theta}_n \notin N_{\theta_0}) \to 0 \quad \text{as} \quad n \to \infty.$$

Proof. Let $\{A_i\}$ be a sequence of neighborhoods which converge monotonically to $\{\varphi\}$ for some $\varphi \in \Theta^* - \{\theta_0\}$. Then by the monotone convergence theorem and (i), (ii) and (iii), as $i \to \infty$,

$$E_{\theta_0}[\inf_{\varphi' \in N_\varphi \cap A_i} \log[f(X|\theta_0)/f(X|\varphi')]] \to I(\theta_0, \varphi) > 0.$$

It follows that for each $\varphi \in \Theta^* - \{\theta_0\}$, there is a neighborhood N_φ^* such that

$$E_{\theta_0}[\inf_{\varphi' \in N_\varphi^*} \log[f(X|\theta_0)/f(X|\varphi')]] > 0.$$

The union of N_{θ_0} with these N_φ^* covers Θ^*. By compactness, a finite subset of these N_φ^* covers $\Theta^* - N_{\theta_0}$. By compactness and continuity $\hat{\theta}_n$ exists (not necessarily uniquely). If $\hat{\theta}_n \notin N_{\theta_0}$, there is one of the finite number of N_φ^* for which

$$\sup_{\varphi' \in N_\varphi^*} \prod_{i=1}^{n} f(X_i|\varphi') \geq \prod_{i=1}^{n} f(X_i|\theta_0)$$

and $\sum_{i=1}^{n} Y_i \leq 0$, where

$$Y_i = \inf_{\varphi' \in N_\varphi^*} \log[f(X_i|\theta_0)/f(X_i|\varphi')]$$

has positive mean. As in the proof for finite Θ, the law of large numbers applies and the theorem follows.

Example 9.4.1. Let $\theta = (\mu_1, \mu_2)$, $-\infty < \mu_1 < \infty$, $-\infty < \mu_2 < \infty$, and

$$f(x, 0|\theta) = \frac{1}{2\sqrt{2\pi}} e^{-(x-\mu_1)^2/2},$$

$$f(x, 1|\theta) = \frac{1}{2\sqrt{2\pi}} e^{-(x-\mu_2)^2/2}.$$

This corresponds to an experiment where a coin is tossed. If a tail falls we observe a normal r.v. with mean μ_1 and if a head falls, we observe a normal r.v. with mean μ_2. The coin is observed, and its outcome is represented by a one for head and 0 for tail. The density is measured with respect to Lebesgue measure on two parallel lines in the (x, x^*)-space. As $\mu_1 \to \infty$, $f(x, x^*|\theta)$ converges to the density of a subdistribution with total probability 0.5. The two-dimensional θ space is compactified by adding two lines at ∞ and a point (∞, ∞). It could also be compactified by adding four lines and four points at (∞, ∞), $(\infty, -\infty)$, etc. Then (μ_1, ∞) and $(\mu_1, -\infty)$ would correspond to the same subdistribution and could not be "separated," in the sense that $I((\mu_1, \infty), (\mu_1, -\infty)) = 0$. But these are not original points of Θ and do not correspond to possible values of θ_0.

It is not at all surprising that consistency applies in this example, but the example is useful in showing that it is useful to extend the original result, due to Wald [W1], which corresponded essentially to adjoining a single point at ∞ with subdistribution $f(x|\theta) = 0$.

Remark 1. The theorem as stated here concerns the maximum-likelihood estimator on Θ^*. The statistician typically does not know Θ^*. He seeks the m.l.e. on Θ which may fail to exist. If there are no points of $\Theta^* - \Theta$ in a neighborhood of θ_0, this problem occurs only with probability approaching zero. Otherwise, we may use the following device. Seek an ε-*m.l.e. estimate* where this is one whose likelihood exceeds $(1 - \varepsilon)$ times the supremum. Such an estimate always exists. Our proof readily extends to imply that with probability one all such estimates are within a neighborhood of θ_0 for n sufficiently large.

Remark 2. If stronger conditions, such as the existence of appropriate m.g.f.'s, are imposed, the large deviation theory can be applied to show that the probability of $\hat{\theta}_n$ lying outside N_{θ_0} approaches zero at an exponential rate. This requires a minor modification of the proof where the law of large numbers was applied.

Remark 3. Consider θ as simply an abstract label attached to the distribution P_θ. From the statistician's point of view, a natural topology tends to be determined by the loss function $L(\theta_0, T)$ which measures the loss incurred by acting as though T is the value of θ when it is actually θ_0. For the theorem prover, the tendency is to consider measures of distance between P_θ and P_φ as determinants of the natural topology. It is of limited use to prove consistency in the wrong topology.

10. Optimal sample size in testing. One aspect of the problem of design of experiments is that of deciding how much experimentation to do. The appropriate amount of experimentation depends on what the true state of nature is. In the large sample size situation where the sample size is specified in advance, one must compromise among the various possible states of nature.

Thus if the cost of sampling is c per observation and the error probabilities of a good test are roughly $\exp(-nI)$ for each state of nature, then one may minimize

$$R = cn + \exp(-nI)$$

by selecting n to be approximately $-I^{-1}\log(c/I)$, and the resulting minimum value of R is asymptotically $I^{-1}(-c\log c)$ as $c \to 0$. This discussion is rough and neglects losses but indicates that for a well-designed experiment for testing $H_1: \theta = \theta_1$ vs. $H_2: \theta = \theta_2$, where the cost per unit of information is small, we should have a large sample, and the cost of experimentation should be the main part of the overall risk.

In this analysis the cost of experimentation is of the order of $-c\log c$ as compared to the risk due to making the wrong decision which is of order of c. The ratio approaches ∞ as $c \to 0$ but in most reasonable problems, this approach to zero is so slow that enormously low error probabilities are required to make the approximations useful for computation. Nevertheless, the role of I in the leading term is instructive and puts the information in a reasonable perspective as an important criterion for judging an experiment.

Up to now (and for some time in the following) we have confined attention to simple testing problems where the alternative hypotheses specify discrete (separated) states of nature. Such problems are somewhat artificial and problems of the type "Test $H_1: \theta \geq \theta_0$ vs. $H_2: \theta < \theta_0$" are often more meaningful. In these cases the appropriate analysis often leads to the Fisher measure of information and indicates a different relationship between the cost of sampling and the risk of making the wrong decision. Both costs are of the same order of magnitude and the appropriate sample size is of the order of $c^{-1/2}$.

Rather than go into a detailed analysis at this point, we simply illustrate with an example in which the calculations are relatively easy.

Example 10.1. The random variable X has the $\mathcal{N}(\mu, \sigma^2)$ distribution with known σ^2. It is desired to test $H_1: \mu \geq 0$ vs. $H_2: \mu < 0$. One may select n independent observations on X at a cost of c per unit observation. The cost of deciding wrong is given by $r(\mu) = k|\mu|$. The unknown mean μ has the prior distribution $\mathcal{N}(\mu_0, \sigma_0^2)$. What is the best sample size and corresponding test?

For a given sample size n, the appropriate test is of the form "*accept* H_1 *if* $\overline{X} \geq k_n$," which has probability of rejection

$$\alpha_n(\mu) = P_\mu(\overline{X} \leq k_n) = \Phi[\sqrt{n}(k_n - \mu)/\sigma],$$

and the (Bayes) risk associated with this procedure is

$$\mathcal{B} = \int_0^\infty k\mu \frac{1}{\sigma_0} \varphi\left[\frac{\mu - \mu_0}{\sigma_0}\right] \Phi[\sqrt{n}(k_n - \mu)/\sigma]\, d\mu$$

$$- \int_{-\infty}^0 k\mu \frac{1}{\sigma_0} \varphi\left[\frac{\mu - \mu_0}{\sigma_0}\right] \Phi[-\sqrt{n}(k_n - \mu)/\sigma]\, d\mu,$$

where φ and Φ are the standard normal density and c.d.f. Rather than get involved at this point in the details of the analysis, let us see what the orders of magnitude are in a simple case. Take $\mu_0 = 0$ and $\sigma_0 = 1$. Then $k_n = 0$ is the natural choice by symmetry and we have

$$\mathscr{B} = 2k \int_0^\infty \mu\varphi(\mu)\Phi[-\sqrt{n}\mu/\sigma]\,d\mu.$$

As $n \to \infty$, Φ approaches zero very rapidly for fixed positive μ.

$$\mathscr{B} \approx \frac{2k}{\sqrt{n}}\sigma\varphi(0)\int_0^\infty \frac{\sqrt{n}\mu}{\sigma}\Phi\left[\frac{-\sqrt{n}\mu}{\sigma}\right]d\mu$$

$$\approx \frac{2k\sigma^2}{n}\varphi(0)\int_0^\infty v\Phi(-v)\,dv = \frac{k\sigma^2\varphi(0)}{2n}.$$

Adding the cost of sampling, we minimize

$$cn + \frac{k\sigma^2\varphi(0)}{2n}$$

and the appropriate sample size is

$$n = k^{1/2}c^{-1/2}\sigma\sqrt{\varphi(0)/2}.$$

The cost of sampling and the risk due to the probability of error are then both equal to

$$c^{1/2}k^{1/2}\sigma\sqrt{\varphi(0)/2}.$$

This cost is considerably greater than that in the problem involving simple hypotheses. Thus the fact that the "hypotheses have common boundaries" has a very substantial effect on the minimum risks attainable.

Raiffa and Schlaiffer [R4], and Antelman [A4] have carried out analyses in this example. Rubin and Sethuraman [R3] have developed an asymptotic theory for various classes of priors, some of which assign positive probability to the boundary between the hypotheses.

11. Sequential probability-ratio test. In principle we should be able to take advantage of the fact that as data arrive, our estimate of the true state of nature becomes more refined and accordingly we may be in a better position to decide whether or not more data are needed. One particularly simple case is that where an inspector has decided to reject a lot of items for purchase if there are as many as 5 defective in a sample of 100. If the first 5 items are defective, there is no need to sample further. The simple procedure of stopping inspection as soon as the decision is inevitable (*curtailed sampling*) could serve as a labor-saving device.

However, suppose that the first three items are defective. This does not imply that there will be two more in the sample of 100, but it certainly casts enough doubt on the quality of the lot, that one may feel justified in stopping inspection immediately and rejecting the lot.

In many practical experimental situations there is little flexibility in adjusting sampling requirements. Nevertheless, a theory which provides a guide as to when sampling should or should not continue is desirable for perspective on how to modify practical procedures for efficiency and on the nature of scientific inference in general.

We discussed the idea of sequential inference in connection with locally optimal designs for estimation and indicated that for estimation problems, the use of locally optimal designs based on the current estimate of θ should be quite efficient. For testing hypotheses, the problem of using current data is a somewhat deeper problem. Wald introduced the sequential probability ratio test to test $H_1 : \theta = \theta_1$ vs. $H_2 : \theta = \theta_2$ on the basis of independent observations on X which has density $f(x|\theta)$.

Wald proposed and evaluated the properties of the following extension of the likelihood-ratio test called the *sequential probability-ratio test* (SPRT). Let $Z_n = (X_1, X_2, \cdots, X_n)$, $\lambda_n(Z_n) = \prod_{i=1}^n f(X_i|\theta_1)/f(X_i|\theta_2)$ and let $A > 1 > B$. For each n, after the nth observation:

Stop sampling and accept H_1 if $\lambda_n(Z_n) \geq A$.

Stop sampling and accept H_2 if $\lambda_n(Z_n) \leq B$.

Take another observation   if $B < \lambda_n(Z_n) < A$.

It is reasonable that the decision rule should be based on λ once sampling is stopped. That sampling should be stopped according to the above rule with fixed A and B is less obvious. However, as we shall see, Bayesian considerations support this procedure and can be used to establish the surprisingly strong optimality result: *For given error probabilities ε_1 and ε_2, this rule minimizes the expected sample size under both H_1 and H_2.* Here two quantities are simultaneously minimized.

A few approximations and results due to Wald are instructive. We assume hereafter that $P_{\theta_1} \neq P_{\theta_2}$.

For notation, let

$$Y = \log[f(X|\theta_1)/f(X|\theta_2)],$$
(11.1)
$$Y_n = \log[f(X_n|\theta_1)/f(X_n|\theta_2)],$$
$$S_n = \sum_{i=1}^n Y_i = \log \lambda_n(Z_n),$$

and let the stopping time be represented by N.

THEOREM 11.1. *The SPRT leads to termination ($N < \infty$) with probability one.*

Proof. Since $P_{\theta_1} \neq P_{\theta_2}$, the distribution of Y_n under P_{θ_1} does not assign probability one to $Y_n = 0$. Thus there is an $\varepsilon > 0$ such that $P_{\theta_1}\{|Y_n| > \varepsilon\} > 0$. Thus for $n_0 > \varepsilon^{-1} \log(A/B)$, $P_{\theta_1}(|S_{n_0}| \geq \log(A/B)) = p$ for some $p > 0$. But $N = \infty$ implies the following sequence of independent events:

$$|S_{n_0}| < \log(A/B), \quad |S_{2n_0} - S_{n_0}| < \log(A/B), \quad |S_{3n_0} - S_{2n_0}| < \log(A/B), \cdots.$$

The probability of all of these occurring is 0.

THEOREM 11.2. *For the SPRT,* $\varepsilon_1 \leqq B(1 - \varepsilon_2)$ *and* $\varepsilon_2 \leqq A^{-1}(1 - \varepsilon_1)$.

Proof. In the infinite-dimensional $(X_1, X_2, \cdots)$-space the region of acceptance of H_2 is the union of the disjoint sets $E_n = \{N = n, \lambda_n \leqq B\}$, $n \geqq 1$:

$$\varepsilon_1 = \varepsilon(\theta_1) = \sum_{n=1}^{\infty} \int_{E_n} \prod_{i=1}^{n} f(x_i|\theta_1)\mu(dx_i)$$

$$= \sum_{n=1}^{\infty} \int_{E_n} \lambda_n(z_n) \prod_{i=1}^{n} f(x_i|\theta_2)\mu(dx_i)$$

$$\leqq B \sum_{n=1}^{\infty} \int_{E_n} \prod_{i=1}^{n} f(x_i|\theta_2)\mu(dx_i) = BP_{\theta_2}(\text{accept } H_2) = B(1 - \varepsilon_2).$$

The second inequality follows by a similar argument.

THEOREM 11.3. *For the SPRT,*

(11.2)
$$E_{\theta_1}(N) \approx [\varepsilon_1 \log B + (1 - \varepsilon_1) \log A]/E_{\theta_1}(Y),$$
$$E_{\theta_2}(N) \approx [(1 - \varepsilon_2) \log B + \varepsilon_2 \log A]/E_{\theta_2}(Y).$$

To establish this almost theorem, we first present the Wald lemma, prefaced by a bit of terminology. Let $Y_1, Y_2, \cdots$ be independent observations on a random variable Y, let $S_n = Y_1 + Y_2 + \cdots + Y_n$, and let N be a *stopping rule*. Informally a stopping rule is a finite integer-valued random variable which takes on the value $n > 0$ depending on the values of $Y_1, Y_2, \cdots, Y_n$ and not on "future" values. More formally, $\{N = n\} \in \mathscr{B}(Y_1, Y_2, \cdots, Y_n)$, the Borel field generated by $Y_1, Y_2, \cdots, Y_n$. That is the class of sets in the original measure space for which $(Y_1, Y_2, \cdots, Y_n)$ falls in a Borel set.

LEMMA 11.1. Wald lemma. *If EY and EN exist finite, then*

(11.3)
$$E(S_N) = E(Y)E(N).$$

This equation also holds if $E(S_N)$ exists so long as it is not the case that both $EN = \infty$ and $EY = 0$ or fails to exist.

Proof. Basically this is a martingale result involving the martingale $\sum_{i=1}^{n} [Y_i - E(Y)]$. A direct and elegant proof in Feller [F1] is the following. Let $N_1, N_2, \cdots$ be the successive values of N observed using $Y_1, Y_2, \cdots, Y_{N_1}, Y_{N_1+1}, \cdots, Y_{N_1+N_2}, \cdots$. Then

$$\frac{Y_1 + \cdots + Y_{N_1} + Y_{N_1+1} + \cdots + Y_{N_1+\cdots+N_r}}{r}$$

$$= \frac{Y_1 + \cdots + Y_{N_1+\cdots+N_r}}{N_1 + \cdots + N_r} \cdot \frac{N_1 + \cdots + N_r}{r}.$$

By the law of large numbers, the right-hand side converges w.p.l. to $E(Y)E(N)$ as $r \to \infty$. The left-hand side is an average of r independent observations of S_N. Its convergence implies that the limit is ES_N. The lemma follows.

Now we easily "prove" Theorem 11.3. Since $E_{\theta_1}(Y) = I(\theta_1, \theta_2) > 0$, it is easy to see that $E_{\theta_1}(S_N) > -\infty$. By Theorem 11.1, N is a stopping rule and $E_{\theta_1}(N) = E_{\theta_1}(S_N)/E_{\theta_1}(Y)$. Upon termination of sampling with the SPRT, $S_N = \log \lambda_N$

is either greater than $\log A$ or less than $\log B$ and in most applications the *excess*, $S_n - \log A$ upon acceptance of H_1 or $\log B - S_n$ upon acceptance of H_2, is relatively small. In such cases,

$$E_{\theta_1} S_N \approx \varepsilon_1 \log B + (1 - \varepsilon_1) \log A.$$

Similarly,

$$E_{\theta_2} S_N \approx (1 - \varepsilon_2) \log B + \varepsilon_2 \log A,$$

and the theorem follows. Moreover, if the excess can be neglected, the bounds in Theorem 11.2 become approximations yielding

(11.4) $\varepsilon_1 \approx B(A - 1)/(A - B)$ and $\varepsilon_2 \approx (1 - B)/(A - B)$.

Recall that $E_{\theta_1}(Y) = I(\theta_1, \theta_2) > 0$ and $E_{\theta_2}(Y) = -I(\theta_2, \theta_1) < 0$.

Theorem 11.3 and Theorem 11.2 can be strengthened by deriving bounds on the effect of the excess. Such bounds appear in Wald [W2].

Until now we have pretty much neglected problems involving more than two possible states of nature. Because the issues and computations in sequential problems are more delicate than in fixed sample size problems, we digress for a brief discussion of the problem of testing $H_1 : \theta \geq \theta_0$ vs. $H_2 : \theta < \theta_0$.

Incidentally, conventional statistical terminology refers to a hypothesis as *simple* if it determines the distribution of the data uniquely and as *composite* otherwise. Thus $H_3 : \theta = \theta_0$ is a simple hypothesis whereas H_1 is composite in nondegenerate cases. The reader should be aware that our considerable emphasis on simple hypotheses needs supplementation to make the results relevant to practical problems where composite hypotheses are much more common.

If it is desired to test $H_1 : \theta \geq \theta_0$ vs. $H_2 : \theta < \theta_0$, Wald suggested that ordinarily the loss for the wrong decision would depend on how far θ is from θ_0. For θ close to θ_0 an incorrect decision is not serious. One may construct an "indifference zone" (θ_2, θ_1) in which we do not care what happens to $\varepsilon(\theta)$. Then one attempts to keep $\varepsilon(\theta)$ small outside. In particular, specifying the values of $\varepsilon(\theta_1) = \varepsilon_1$ and $\varepsilon(\theta_2) = \varepsilon_2$, one ordinarily anticipates that for $\theta > \theta_1$, $\varepsilon(\theta)$ will decrease below ε_1 and for $\theta < \theta_2$, ε will decrease below ε_2. Thus Wald suggested that the composite hypothesis problem be treated as that of testing the simple hypothesis $H_1^* : \theta = \theta_1$ vs. $H_2^* : \theta = \theta_2$ for appropriately selected θ_1 and θ_2.

This approach is a reasonable one for many problems. However, as a basis for asymptotic theoretical development it leads to substantial difficulties as we shall see later. Wald did suggest another approach, more Bayesian in nature, but that is difficult to apply analytically and not much was done with it.

Suppose now that we apply the SPRT for H_1^* vs. H_2^* and use the resulting decision in the problem of testing H_1 vs. H_2. By this I mean we observe data, $X_1, X_2, \cdots$ from $f(X|\theta)$, θ unknown,

$$\lambda_n = \prod_{i=1}^{n} [f(X_i|\theta_1)/f(X_i|\theta_2)]$$

and when the SPRT calls for stopping and accepting H_1^*, we stop and accept H_1, etc. This is a well-defined procedure. To evaluate this we need to know $E_\theta(N)$ and $\varepsilon(\theta)$ for arbitrary θ, and not only for θ_1 and θ_2. The theorems above extend to this problem.

In the following statements any probability and expectation refers to P_θ unless specified otherwise.

THEOREM 11.1*. *The SPRT leads to termination with probability 1 unless* $Y = \log[f(X|\theta_1)/f(X|\theta_2)] = 0$ *with* P_θ *probability 1.*

Proof. The proof of Theorem 11.1 applies unchanged.

THEOREM 11.2*. *If* $M_Y[h(\theta)] = 1$ *for* $h(\theta) \neq 0$, *and* $E_\theta(Y) \neq 0$, *then*

$$P_\theta(\text{reject } H_1) \approx \frac{1 - A^{h(\theta)}}{B^{h(\theta)} - A^{h(\theta)}}.$$

Proof. Let

$$g(x|\varphi_1) = f(x|\theta)[f(x|\theta_1)/f(x|\theta_2)]^{h(\theta)}$$

and

$$g(x|\varphi_2) = f(x|\theta),$$

where $h \neq 0$ is chosen so that $g(x|\varphi_1)$ is a density, i.e., so that

$$M_Y[h(\theta)] = E_\theta[e^{h(\theta)Y}] = 1.$$

The SPRT of $H_1^{**}: \varphi = \varphi_1$ vs. $H_2^{**}: \varphi = \varphi_2$ using $A^{**} = A^{h(\theta)}$ and $B^{**} = B^{h(\theta)}$ is exactly the same as our procedure (if $h > 0$). In that case $P_\theta(\text{reject } H_1) = 1 - \varepsilon_2^{**}$. Applying (11.4), which is based on neglecting the excess in Theorem 11.2, $\varepsilon_2^{**} \approx (1 - B^{h(\theta)})/(A^{h(\theta)} - B^{h(\theta)})$. Thus

$$P_\theta(\text{reject } H_1) \approx \frac{1 - A^{h(\theta)}}{B^{h(\theta)} - A^{h(\theta)}} \quad \text{if} \quad h(\theta) > 0.$$

The case $h(\theta) < 0$ follows similarly, after identifying ε_2^{**} of the SPRT of H_1^{**} vs. H_2^{**}, using $A^{**} = B^{h(\theta)}$ and $B^{**} = A^{h(\theta)}$, with $P_\theta(\text{reject } H_1)$ of the SPRT of H_1 vs H_2.

Ordinarily there is an $h(\theta) \neq 0$ for which the condition of Theorem 11.2* tends to be satisfied. Since $M_Y(t)$ is convex and $M_Y'(0) = E_\theta(Y)$, as long as $E_\theta(Y) \neq 0$, $M_Y(t)$ tends to go below one for small values of t with sign opposite to that of $E_\theta(Y)$. If $P_\theta(Y < 0)$ and $P_\theta(Y > 0)$ are both positive, $M_Y(t)$ will go to ∞ as $t \to \pm\infty$. Thus there will be some value of $h(\theta)$ with sign opposite to that of $E_\theta(Y)$ for which $M_Y(h(\theta)) = 1$, unless $M_Y(t)$ jumps from a value below one to ∞ at the end of its domain of finite existence.

THEOREM 11.3*. *If* $E_\theta(Y) \neq 0$,

$$(11.5) \qquad E_\theta(N) \approx [E_\theta(Y)]^{-1}[P_\theta\{\text{reject } H_1\} \log B + P_\theta\{\text{accept } H_1\} \log A].$$

Proof. This follows immediately from the Wald lemma.

We illustrate these theorems with the normal example.

Example 11.1. Use the SPRT of $H_1^*:\mu = 1$ vs. $H_2^*:\mu = -1$ to test $H_1:\mu \geqq 0$ vs. $H_2:\mu < 0$, where $X_1, X_2, \cdots$ are independent observations on X with the $\mathcal{N}(\mu, \sigma^2)$ distribution, σ^2 known.

Here

$$Y = \log[f(X|\mu_1)/f(X|\mu_2)] = 2X\sigma^{-2}.$$

If we use the SPRT of H_1^* vs. H_2^* so that $\varepsilon_1^* \approx \varepsilon_2^* \approx 0.05$, Theorem 11.2 suggests that we use $B = \varepsilon_1/(1 - \varepsilon_2) \approx 0.05$ and $A = (1 - \varepsilon_1)/\varepsilon_2 \approx 20$. Then $-\log B = \log A = 2.995$.

The SPRT calls for:

$$\text{Stop and accept } H_1 \quad \text{if} \quad \sum X_i > 1.50\sigma^2.$$

$$\text{Stop and reject } H_1 \quad \text{if} \quad \sum X_i < -1.50\sigma^2.$$

$$\text{Continue sampling} \quad \text{otherwise.}$$

For general μ,

$$E(Y) = 2\sigma^{-2}\mu,$$

$$M_Y(t) = \exp[2\mu\sigma^{-2}t + 2\sigma^{-2}t^2].$$

Thus

$$h(\mu) = -\mu.$$

Then for $\mu \neq 0$,

$$P\{\text{reject } H_1\} \approx \frac{1 - (0.05)^\mu}{20^\mu - (0.05)^\mu} = \alpha(\mu)$$

and

$$E(N) \approx \frac{\sigma^2}{2\mu}[1 - 2\alpha(\mu)](2.995).$$

For $\mu = 0$, a more delicate approach is needed.

A variation of the proof of Theorems 11.2 and 11.2* comes from the following fundamental identity. The use of this identity provides a means of obtaining upper and lower bounds on $\alpha(\theta) = P_\theta(\text{reject } H_1)$ as well as the possibility of treating the case where $E_\theta(Y) = 0$.

LEMMA 11.2 (Fundamental identity).

$$E(e^{S_N t} \cdot M_Y^{-N}(t)) = 1 \quad \text{if} \quad 1 \leqq |M_Y(t)| < \infty.$$

We shall not prove this identity which can be derived by regarding $\exp(S_n t)M_Y^{-n}(t)$ as a martingale. Letting $t = h(\theta)$, we have $E[\exp(h(\theta)S_N)] = 1$, which converts to

$$\alpha(\theta)E_2 + [1 - \alpha(\theta)]E_1 = 1,$$

where $E_1 = E[\exp(h(\theta)S_N)|S_N > \log A] \approx A^{h(\theta)}$ and $E_2 = E[\exp(h(\theta)S_N)|S_N < \log B]$ $\approx B^{h(\theta)}$. Theorem 11.2* follows immediately from these approximations. Moreover, both E_1 and E_2 can be bounded in special examples. Bounds may be easily derived in terms of

$$\sup_a a\, E\, (e^{h(\theta)Y}|e^{h(\theta)Y} \leq a^{-1})$$

and

$$\inf_b b\, E\, (e^{h(\theta)Y}|e^{h(\theta)Y} \geq b^{-1}).$$

12. Optimality of sequential probability-ratio test. The SPRT has the following rather extraordinary optimality property.

THEOREM 12.1. *If an SPRT has error probabilities ε_1 and ε_2 and expected sample sizes $E_{\theta_1}(N)$ and $E_{\theta_2}(N)$, any other sequential test with error probabilities $\varepsilon'_1 \leq \varepsilon_1$ and $\varepsilon'_2 \leq \varepsilon_2$ and finite expected sample sizes $E_{\theta_1}(N')$ will satisfy $E_{\theta_1}(N') \geq E_{\theta_1}(N)$ and $E_{\theta_2}(N') \geq E_{\theta_2}(N)$.*

A rigorous proof involves a number of technicalities. To avoid them we present a heuristic outline of the main ideas. In outline, we first "show" that the Bayes procedure is an SPRT for the problem where the losses for doing the wrong thing are $r_1 = r(\theta_1)$ and $r_2 = r(\theta_2)$, the cost per observation is c and the prior probabilities are $\xi_1 = 1 - \xi$ and $\xi_2 = \xi$. Next we suggest that a fixed SPRT is a Bayes procedure for each of a class of problems where $\xi \to 0$ and where $\xi \to 1$ with r_1, r_2 and c changing with ξ. The optimality will follow.

Part I. *The SPRT as a Bayes procedure.* Given any sequential procedure s, let

$$R(\theta, s) = cE_\theta(N) + r(\theta)\varepsilon(\theta), \quad \theta = \theta_1, \theta_2.$$

We seek to find the Bayes procedure, i.e., that sequential procedure which minimizes the *Bayes risk*

(12.1) $$\mathscr{R}(\xi, s) = (1 - \xi)R(\theta_1, s) + \xi R(\theta_2, s).$$

Each procedure may be represented by a straight line in the $(\xi, \mathscr{R})$-space (see Fig. 7). Let

(12.2) $$\mathscr{R}(\xi) = \inf_s \mathscr{R}(\xi, s),$$

which is a *concave function* indicating the best that can be achieved for the prior probabilities $(1 - \xi, \xi)$. (The infimum of a sequence of concave functions is concave and straight lines are concave.) Let $\mathscr{R}_1(\xi)$ be the optimum achieved by procedures which involve at least one observation. Let $\mathscr{R}_0(\xi)$ be the optimum achieved by procedures which take no observations. Then

(12.3) $$\mathscr{R}_0(\xi) = \min[r_2\xi, r_1(1 - \xi)].$$

$\mathscr{R}_1(\xi) \geq c$ is concave for the same reason that $\mathscr{R}(\xi)$ was. It is clear that either $\mathscr{R}_0$ and $\mathscr{R}_1$ intersect at two points ξ_1 and $\xi_2 > \xi_1$, or $\mathscr{R}_1(\xi) \geq \mathscr{R}_0(\xi)$ for all ξ (see Fig. 8). In the latter case, the Bayes procedure always calls for a decision without

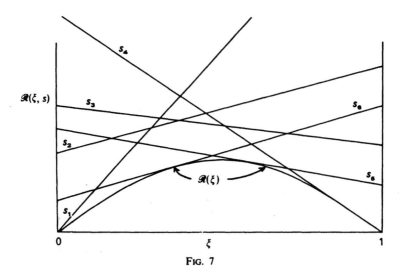

FIG. 7

data (the cost of sampling always exceeds the risk of deciding). In the former case, an optimal procedure (*assuming it exists*) would call for accepting H_1 without observation if $\xi \leq \xi_1$, rejecting H_1 without observation if $\xi \geq \xi_2$ and taking an observation if $\xi_1 < \xi < \xi_2$. Moreover,

$$(12.4) \qquad \mathscr{R}(\xi) = \min[R_0(\xi), R_1(\xi)].$$

If an observation were taken, the prior probability ξ would be replaced by the posterior probability ξ_X. Otherwise, the problem confronting the statistician (who

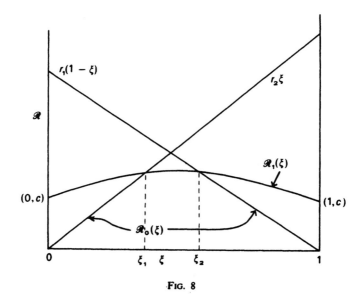

FIG. 8

has already paid c units for the first observation) is unaltered and once more he would decide to stop if $\zeta_X \leq \zeta_1$ or if $\zeta_X \geq \zeta_2$ and continue sampling otherwise.

Since the posterior probability based on $X_1, X_2, \cdots, X_n$ is

$$\zeta_{(n)} = \frac{1}{1 + ((1 - \zeta)/\zeta)\lambda_n(Z_n)},$$

which is monotone in λ_n, the Bayes procedure is:

Stop and accept H_1 if $\lambda_n(Z_n) \geq \dfrac{1 - \zeta_1}{\zeta_1} \dfrac{\zeta}{1 - \zeta} = A(\zeta, \zeta_1).$

Stop and reject H_1 if $\lambda_n(Z_n) \leq \dfrac{1 - \zeta_2}{\zeta_2} \dfrac{\zeta}{1 - \zeta} = B(\zeta, \zeta_2).$

Continue sampling otherwise.

Part II. Varying costs and priors for fixed SPRT. The limits $A(\zeta, \zeta_1)$ and $B(\zeta, \zeta_2)$ depend on ζ, r_1, r_2 and c since ζ_1 and ζ_2 depend on r_1, r_2, c. We seek a set of ζ, r_1, r_2 and c for which $\zeta \to 0$ and $A(\zeta, \zeta_1)$ and $B(\zeta, \zeta_2)$ assume fixed specified values A and B. We need appropriate ζ_1 and ζ_2 approaching zero. There are three cost parameters to adjust but effectively just the proportions r_1/c and r_2/c count. Let $r_2/c \to \infty$ and r_1/c decrease toward 1. Then it is clear (see Fig. 8) that ζ_1 and ζ_2 are achievable. Matthes [M1] has a proof which uses the continuity of ζ_1 and ζ_2 as functions of (r_1, r_2, c) as well as a theorem of Rado and Reichelderfer. Thus a given SPRT is a Bayes procedure for a sequence of ζ, r_1, r_2, c, where $\zeta \to 0$.

Part III. Optimality of SPRT. Let s' be an alternate procedure with error probabilities $\varepsilon_1' \leq \varepsilon_1$ and $\varepsilon_2' \leq \varepsilon_2$ and finite expected sample sizes $E_{\theta_1}(N')$ and $E_{\theta_2}(N')$. Then for the appropriately selected $\zeta \to 0, r_1, r_2, c,$

$$(1 - \zeta)[r_1\varepsilon_1 + cE_{\theta_1}(N)] + \zeta[r_2\varepsilon_2 + cE_{\theta_2}(N)] \leq (1 - \zeta)[r_1\varepsilon_1' + cE_{\theta_1}(N')]$$
$$+ \zeta[r_2\varepsilon_2' + cE_{\theta_2}(N')]$$

and hence,

$$(1 - \zeta)E_{\theta_1}(N) + \zeta E_{\theta_2}(N) \leq (1 - \zeta)E_{\theta_1}(N') + \zeta E_{\theta_2}(N').$$

Taking limits as $\zeta \to 0$,

$$E_{\theta_1}(N) \leq E_{\theta_1}(N').$$

A symmetric argument (with $\zeta \to 1$) yields

$$E_{\theta_2}(N) \leq E_{\theta_2}(N'),$$

which is the desired result.

A rigorous argument in Part I requires a proof of the existence of an optimal procedure. Arrow, Blackwell and Girshick developed such a proof using the idea of backward induction with a specified maximum number of observations m, and letting $m \to \infty$. The idea of using Bayes strategies appeared in the original proof of Wald and Wolfowitz [W3].

13. Motivation for an approach to sequential design of experiments in testing hypotheses. Our basic approach will consist of studying the behavior and efficiency of the Bayes sequential procedure as $c \to 0$, which leads to increasing sample size. We test $H_1 : \theta = \theta_1$ vs. $H_2 : \theta = \theta_2$ as in the previous sections. Assuming the approximations previously derived are valid we shall motivate general results that can be proved rigorously.

As $c \to 0$, one expects $A \to \infty$ and $B \to 0$. Assuming that the "excess" can be neglected,

$$(13.1) \qquad\qquad \varepsilon_1 \approx B \quad \text{and} \quad \varepsilon_2 \approx A^{-1}.$$

Assuming ε_1 and ε_2 are comparable,

$$(13.2) \qquad\qquad \begin{aligned} E_{\theta_1}(N) &\approx \log A / I_1, \\ E_{\theta_2}(N) &\approx -\log B / I_2, \end{aligned}$$

where $I_1 = I(\theta_1, \theta_2) = E_{\theta_1}(Y)$ and $I_2 = I(\theta_2, \theta_1) = -E_{\theta_2}(Y)$ are the Kullback–Leibler information numbers. The last two equations indicate that under H_i, $S_n = \sum_{j=1}^{n} Y_j$ tends to increase at a rate of I_i per observation and the expected sample size is roughly the number of observations needed for S_n, moving at this rate, to reach the boundary for the correct decision. Then for the SPRT test s determined by A and B,

$$\mathcal{R}(\xi, s) \approx (1 - \xi)\left[r_1 B + \frac{c \log A}{I_1} \right] + \xi \left[r_2 A^{-1} - \frac{c \log B}{I_2} \right].$$

To minimize with respect to A and B, we set the partial derivatives equal to zero:

$$\frac{\partial \mathcal{R}}{\partial A} = (1 - \xi)\frac{c}{A I_1} - \xi r_2 A^{-2} = 0,$$

$$\frac{\partial \mathcal{R}}{\partial B} = (1 - \xi)r_1 - \frac{\xi c}{B I_2} = 0.$$

Hence we have for optimal A and B,

$$A \approx \frac{\xi}{1 - \xi} \frac{r_2 I_1}{c}, \quad B \approx \frac{\xi}{1 - \xi} \frac{c}{r_1 I_2},$$

$$(13.3) \qquad\qquad \varepsilon_1 \sim c, \quad \varepsilon_2 \sim c,$$

$$E_{\theta_1} N \approx -\frac{\log c}{I_1}, \quad E_{\theta_2}(N) \approx -\frac{\log c}{I_2},$$

$$R(\theta_1, s) = cE_{\theta_1}(N) + r_1 \varepsilon_1 \approx -\frac{c \log c}{I_1},$$

$$(13.4) \qquad\qquad R(\theta_2, s) = cE_{\theta_2}(N) + r_2 \varepsilon_2 \approx -\frac{c \log c}{I_2}.$$

Suppose now that the experimenter had a choice between two sequential experiments each costing c per observation. In the first, the random variable X is used with Kullback–Leibler information numbers I_1 and I_2. In the second, X^* is used with information numbers I_1^* and I_2^*. Once the choice is made, no switching is permitted.

If $I_1 > I_1^*$ and $I_2 > I_2^*$, the above equation suggests that X should be used. Suppose $I_1 > I_1^*$ and $I_2 < I_2^*$. Then X should be used if $H_1 : \theta = \theta_1$ is correct and X^* should be used if $H_2 : \theta = \theta_2$ is correct. If the experimental designer knew the true state of nature, he could select the better experiment to be analyzed. Of course, if he knew the true state of nature, he would not be interested in the testing problem.

In the more realistic context where the statistician can shift experiments as he accumulates information he may not know θ but he may reach a point where he is pretty sure of θ. In view of the low cost of sampling he may not be sure enough to stop sampling and make a definitive terminal decision, but sure enough to have a preference among the available experiments.

This reasoning suggests that as a coarse first approximation it makes sense to use the following procedure.

Let λ_n be the likelihood-ratio based on the first n observations.

Stop sampling and select H_1 if $\lambda_n > c^{-1}$.

Stop sampling and select H_2 if $\lambda_n < c$.

If $c < \lambda_n < c^{-1}$, select another observation from among the available experiments so as to maximize $I_1 = I(\theta_1, \theta_2)$ if $\lambda_n \geq 1$ and so as to maximize $I_2 = I(\theta_2, \theta_1)$ if $\lambda_n < 1$.

Let us now proceed to the simplest composite problem: $H_1 : \theta = \theta_1$ vs. $H_2 : \theta = \theta_2$ or $\theta = \theta_3$. Let $\xi = (\xi_1, \xi_2, \xi_3)$ represent the prior probability distribution which can be conveniently represented in barycentric coordinates (see Fig. 9). In the no-data problem with two actions, corresponding to accepting H_1 and H_2, the losses for the wrong decision may be represented by $r_1 = r(\theta_1)$, $r_2 = r(\theta_2)$ and

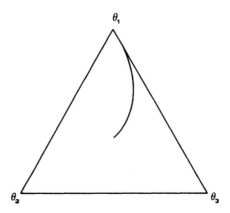

FIG. 9. *Barycentric coordinates.* ξ_i = *distance to side opposite vertex labeled* θ_i *in equilateral triangle with unit height*

$r_3 = r(\theta_3)$. The Bayes risk associated with accepting H_1 is $\xi_2 r_2 + \xi_3 r_3$ and with accepting H_2 is $\xi_1 r_1$. Thus the optimal choice gives Bayes risk

(13.5) $$\mathscr{B}_0(\xi) = \min[\xi_1 r_1, \xi_2 r_2 + \xi_3 r_3].$$

After n independent observations $X_1, X_2, \cdots, X_n$ on a random variable X with density $f(x|\theta)$, the posterior probabilities are, according to Bayes' theorem, proportional to

$$\xi_i L(Z_n|\theta_i) = \xi_i \prod_{j=1}^{n} f(X_j|\theta_i), \qquad\qquad i = 1, 2, 3.$$

Thus

(13.6) $$\frac{\xi_{2n}}{\xi_{1n}} = \frac{\xi_2}{\xi_1} e^{-S_{12n}}$$

and

(13.6') $$\frac{\xi_{3n}}{\xi_{1n}} = \frac{\xi_3}{\xi_1} e^{-S_{13n}},$$

where

(13.7) $$S_{12n} = \sum_{i=1}^{n} \log \frac{f(X_i|\theta_1)}{f(X_i|\theta_2)} = \sum_{i=1}^{n} Y_{12i}$$

and

(13.7') $$S_{13n} = \sum_{i=1}^{n} \log \frac{f(X_i|\theta_1)}{f(X_i|\theta_3)} = \sum_{i=1}^{n} Y_{13i}.$$

Suppose now that $H_1 : \theta = \theta_1$ is true. Then (neglecting the subscript θ_1 in E)

$$E(Y_{12i}) = I(\theta_1, \theta_2) > 0$$

and

$$E(Y_{13i}) = I(\theta_1, \theta_3) > 0,$$

and assuming $\xi_1 \neq 0$, ξ_{2n} and ξ_{3n} approach zero roughly as $\exp[-nI(\theta_1, \theta_2)]$ and $\exp[-nI(\theta_1, \theta_3)]$. The risk of wrong decision if sampling is stopped at the nth observation is $r_2 \xi_{2n} + r_3 \xi_{3n}$ which is roughly like

$$\exp[-nI(\theta_1)],$$

where

(13.8) $$I(\theta_1) = \min[I(\theta_1, \theta_2), I(\theta_1, \theta_3)].$$

If $I(\theta_1, \theta_2) > I(\theta_1, \theta_3)$, the path of $(\xi_{1n}, \xi_{2n}, \xi_{3n})$ is represented graphically by the curve which hugs the side opposite θ_2 as it moves toward the vertex θ_1.

When H_1 is true, the experimenter should select an experiment which maximizes $I(\theta_1)$. Consider the situation described in Table 2 where three experiments are

TABLE 2

	e_1	e_2	e_3
$I(\theta_1, \theta_2)$	6	3	2
$I(\theta_1, \theta_3)$	2	3	6
$I(\theta_1)$	2	3	2

available and H_1 is assumed to be true. Of these three experiments, e_2 maximizes $I(\theta_1)$. On the other hand, the randomized experiment where e_1 and e_3 are selected, each with probability 1/2, has $I(\theta_1, \theta_2) = I(\theta_1, \theta_3) = I(\theta_1) = 4$. This suggests the potential advantage of randomized experiments.

Comparison with the case of testing a simple hypothesis versus a simple alternative suggests stopping when the posterior risk $\mathcal{R}_0(\xi_{(n)})$ goes below c. We now have a number of insights which suggest a general approach. After introducing some formal notation and terminology, a sequential procedure incorporating these insights will be recommended and a theorem stating the optimality of this approach presented.

Let $\mathcal{E}$ be a set e of available elementary experiments whose outcome $X(e)$ has density $f(x|\theta, e)$ with respect to a measure μ_e. Let $H_1 : \theta \in \Theta_1$ and $H_2 : \theta \in \Theta_2$ be two hypotheses, where Θ_1 and Θ_2 are disjoint sets whose union Θ represents the set of possible "states of nature". Let $r(\theta)$ be the cost of deciding incorrectly and let c be the cost per observation. After each observation, a decision is made to stop or continue sampling. If sampling is stopped, either H_1 or H_2 is accepted. If sampling is continued, a new experiment e_{n+1} is selected. While its choice may depend on the past data, $Z_n = (X_1, X_2, \cdots, X_n)$, once the experiment is selected, its outcome is independent of the past.

Let $\mathcal{E}^*$ be the set of randomized experiments generated by $\mathcal{E}$. For $e \in \mathcal{E}^*$, $Y(\theta, \varphi, e) = \log[f(X_e|\theta, e)/f(X_e|\varphi, e)]$ has expectation $I(\theta, \varphi, e)$ and variance $v(\theta, \varphi, e)$. Let $Y_n(\theta, \varphi) = \log[f(X_n|\theta, e_n)/f(X_n|\varphi, e_n)]$ and

$$(13.9) \qquad S_n(\theta, \varphi) = \sum_{i=1}^{n} Y_i(\theta, \varphi)$$

and $\hat{\theta}_n$ be the m.l.e. of θ based on Z_n. Let $h(\theta) = \Theta_i$ and $a(\theta) = \Theta - \Theta_i$ if $\theta \in \Theta_i$, $i = 1, 2$, and let $\tilde{\theta}_n$ be the m.l.e. of θ based on Z_n when θ is restricted to $a(\hat{\theta}_n)$. Here $h(\theta)$ is identified with the "hypothesis of θ" and $a(\theta)$ with the "alternative to the hypothesis of θ", and $\tilde{\theta}_n$ is the m.l.e. on the alternative to the hypothesis of $\hat{\theta}_n$.

Procedure A: Stop sampling and accept the hypothesis $h(\hat{\theta}_n)$ if

$$(13.10) \qquad S_n(\hat{\theta}_n, \tilde{\theta}_n) > -\log c.$$

Otherwise select $e_{(n+1)} \in \mathcal{E}^*$ to maximize

$$(13.11) \qquad \inf_{\varphi \in a(\hat{\theta}_n)} I(\hat{\theta}_n, \varphi, e).$$

On the basis of our motivation one would anticipate that this procedure would be asymptotically optimal, that its risk would be asymptotically

(13.12) $$R(\theta) \approx -c \log c / I(\theta),$$

where

(13.13) $$I(\theta) = \sup_{e \in \mathscr{E}^*} \inf_{\varphi \in \Omega(\theta)} I(\theta, \varphi, e),$$

and that the main contribution to this risk would be due to the expected cost of sampling. This conjecture is valid as long as Θ_1 and Θ_2 are "separated". A serious difficulty develops if Θ_1 and Θ_2 have a natural common boundary or if $I(\theta)$ can approach 0 for $\theta \in \Theta$.

Before we proceed to state and prove a theorem to this effect, let us mention a few potential shortcomings. First, the separation requirement is important and getting around it will require new insights. Second, this approach is crude in that it involves pretending that the current estimate $\hat{\theta}_n$ of θ is correct in deciding what experiment to select next. No attempt is made to distinguish between an imprecise estimate of θ based on little evidence and a very precise estimate. As a result, the behavior of this procedure may be poor for problems where moderate sample sizes are anticipated. Third, the stopping rule invokes $\log c$ which is dimensionally incorrect. A more sensible rule would be to stop when the posterior risk, given some initial prior distribution, is reduced to some number of the order of magnitude of c or to replace $\log c$ by $\log[c/r(\hat{\theta}_n)]$ which would at least be dimensionally correct. Fourth, the main term in the risk is of the order $-c \log c$ which is greater by an order of magnitude than the risk due to making the wrong decision. However, $\log c$ approaches $-\infty$ very slowly, and unless error probabilities are very small, the approximation used for $R(\theta)$ is highly undependable even though it furnishes a good guide for experimentation when θ is estimated accurately. We shall elaborate on some of these comments later.

14. Asymptotic optimality of Procedure A in sequential design.

THEOREM 14.1. *If $\mathscr{E}$ and Θ are finite and for every $e \in \mathscr{E}$, $I(\theta, \varphi, e) > 0$ and $v(\theta, \varphi, e) < \infty$ for all θ and $\varphi \in \Theta$ with $\theta \neq \varphi$, and $I(\theta) > 0$, then the risk $R(\theta)$ for Procedure A satisfies*

(14.1) $$R(\theta) \leq [1 + o(1)] \left\{ \frac{-c \log c}{I(\theta)} \right\} \quad \text{for all} \quad \theta,$$

and any procedure s^* for which the risk $R^*(\theta) = O(-c \log c)$ for all θ satisfies

(14.2) $$R^*(\theta) \geq [1 + o(1)] \left\{ \frac{-c \log c}{I(\theta)} \right\} \quad \text{for all} \quad \theta.$$

Theorem 14.1 is a statement of the asymptotic optimality of A in the following sense. If, for all θ, the risk of a procedure is not greater by an order of magnitude than that of A, it must be asymptotically at least as large for *all* θ.

Proof. We outline parts of the proof from [C8]. For notational convenience let θ_0 represent the "true" state of nature and P and E without subscripts will refer to P_{θ_0}.

First, a conventional argument, such as that used in § 9.4 on consistency of the m.l.e., using the bound on $P(\bar{X} < k)$ of § 9.1, shows that irrespective of the design, $\hat{\theta}_n \to \theta_0$ w.p.1. and that this convergence is at a geometric rate, i.e.,

$$(14.3) \qquad P\{\hat{\theta}_m = \theta_0 \text{ for all } m \geq n\} \geq 1 - Ke^{-bn}, \qquad n = 1, 2, \cdots,$$

for some K and $b > 0$.

Second, once $\hat{\theta}_n = \theta_0$, $S_n(\hat{\theta}_n, \varphi)$ increases at a rate determined by $I(\theta_0, \varphi)$ for each $\varphi \in a(\theta_0)$ and another slightly more complex application of the same bound together with the above result yields

$$(14.4) \qquad E(N) \leq [1 + o(1)](-\log c)/I(\theta_0),$$

where N is the random sample size.

Third, essentially the same proof as that used to derive $\varepsilon_1 \leq B(1 - \varepsilon_2)$ for the SPRT gives

$$(14.5) \qquad P\{\hat{\theta}_N = \varphi\} < c \quad \text{for each} \quad \varphi \in a(\theta_0),$$

where $P\{\hat{\theta}_N = \varphi\}$ is the probability of stopping when $\hat{\theta}_N = \varphi$. Since $a(\theta_0)$ has only a finite number of elements,

$$(14.6) \qquad \varepsilon(\theta_0) = O(c).$$

Combining these three results we have (14.1).

A fourth step is necessary to show that $R^*(\theta) = O(-c \log c)$ for all θ only if the error probabilities are not larger than $O(-c \log c)$ and that this implies that $S_N(\theta_0, \varphi)$, the log likelihood-ratio upon stopping, must be sufficiently large with high probability, i.e.,

$$(14.7) \quad P[S_N(\theta_0, \varphi) < -(1 - \varepsilon) \log c] = O(-c^\varepsilon \log c) \quad \text{for all} \quad \varphi \in a(\theta_0).$$

Let $A_n = \{N = n, S_n(\theta_0, \varphi) < -(1 - \varepsilon) \log c, h(\theta_0) \text{ accepted}\}$. Then

$$P[S_n(\theta_0, \varphi) < -(1 - \varepsilon) \log c] \leq \sum_{n=1}^{\infty} P(A_n) + P[\text{reject } h(\theta_0)].$$

The last term is $O(c \log c)$. The sum is bounded by the following argument. There is a number K such that

$$-Kc \log c \geq P_\varphi\{\text{accept } h(\theta_0)\} \geq \sum_{n=1}^{\infty} \int_{A_n} f_n(z_n|\varphi)\mu_n(dz_n),$$

where $f_n(z_n|\varphi)$ represents the density on the sample space of the first n observations.

Then

$$-Kc \log c \geq \sum_{n=1}^{\infty} \int_{A_n} \frac{f_n(z_n|\varphi)}{f_n(z_n|\theta_0)} f_n(z_n|\theta_0)\mu_n(dz_n)$$

$$= \sum_{n=1}^{\infty} \int_{A_n} e^{-S_n(\theta_0,\varphi)} f_n(z_n|\theta_0)\mu_n(dz_n) \geq c^{1-\varepsilon} \sum_{n=1}^{\infty} P(A_n).$$

The desired inequality follows.

Finally, a fifth step is required to show that $S_N(\theta_0, \varphi)$ can be this large with large probability for all $\varphi \in a(\theta_0)$ only if the expected sample size is sufficiently large. More precisely, we prove

(14.8) $$P\left\{ \max_{1 \leq m \leq n} \min_{\varphi \in a(\theta_0)} \sum_{i=1}^{m} Y_i(\theta_0, \varphi) \geq n[I(\theta_0) + \varepsilon] \right\} \to 0 \quad \text{as} \quad n \to \infty,$$

after which, relating $n[I(\theta_0) + \varepsilon]$ with $-(1 - \varepsilon) \log c$ indicates that n is almost sure to exceed a value asymptotically equal to $-\log c/I(\theta_0)$. Equation (14.8) is derived by decomposing

$$\sum_{i=1}^{m} Y_i(\theta_0, \varphi) = \sum_{i=1}^{m} [Y_i(\theta_0, \varphi) - I(\theta_0, \varphi, e_i)] + \sum_{i=1}^{m} I(\theta_0, \varphi, e_i),$$

where the first sum

$$A_{1m} = \sum_{i=1}^{m} [Y_i(\theta_0, \varphi) - I(\theta_0, \varphi, e_i)]$$

is a martingale and

$$A_{2m} = \sum_{i=1}^{m} I(\theta_0, \varphi, e_i)$$

represents m times the payoff for a game where nature selects φ and the experimenter selects some mixture of his available strategies. This game has value $I(\theta_0)$. Thus

$$\min_{\varphi \in a(\theta_0)} A_{2m} \leq mI(\theta_0) \leq nI(\theta_0).$$

Hence it suffices to show that $\max_{1 \leq m \leq n} A_{1m}$ can exceed $n\varepsilon$ only with small probability. But Doob's extension for martingales of the Kolmogorov inequality [D1] states

$$P\left\{ \max_{1 \leq m \leq n} A_{1m} > n\varepsilon \right\} \leq K/n\varepsilon^2 \quad \text{for each} \quad \varphi \in a(\theta_0),$$

and the proof is completed.

15. Extensions and open questions in sequential design. Theorem 14.1 has been extended in several directions. Some of these extensions are relatively direct. Bessler [B2], [B3] treated the case of k actions and infinite sets $\mathscr{E}$ of available experiments. The k-action case corresponds to the situation where there are several possible terminal choices, each of which is optimal for some values of θ but not for others. For later reference we present a somewhat artificial illustrative example.

Example 15.1. Drug screening. In a drug screening experiment, k drugs are tested each of which has probability p of failure with one exception. That one has probability γp of failure. The parameters p and γ are between 0 and 1 and are known. The object is to find which drug has the "odd" probability γp.

In this problem there are k actions. The state of nature θ_i corresponds to the case where the ith drug is the "odd" one and the experiment e_j to the trial of the jth drug. Then for $\theta_i \neq \theta_j$,

$$I(\theta_i, \theta_j, e_h) = \begin{cases} a & \text{if } i = h, \\ b & \text{if } j = h, \\ 0 & \text{otherwise}, \end{cases}$$

where

$$a = \log \gamma^{\gamma p}[(1 - \gamma p)/(1 - p)]^{1 - \gamma p}$$

and

$$b = \log \gamma^{-p}[(1 - p)/(1 - \gamma p)]^{1 - p}.$$

Suppose $k = 3$ and $\hat{\theta}_n = \theta_1$. Then the statistician must select e to maximize $\min_\varphi I(\theta_1, \varphi, e)$ for $\varphi \in a(\theta_1) = \{\theta_2, \theta_3\}$. Thus he must play the game represented below:

I \ II	e_1	e_2	e_3
θ_2	a	b	0
θ_3	a	0	b

There are two cases as follows.

Case 1. $a \geq b/2$, choose e_1.

Case 2. $a < b/2$, choose e_2 and e_3 each with probability $1/2$.

The value of the game is

$$I(\theta_1) = \max(a, b/2).$$

Albert [A1] extended the results to the two-action problem where Θ may be an infinite set. The results break down if $I(\theta) \to 0$. Thus asymptotic optimality for Procedure A requires "separating" the hypotheses. This raises a serious problem which we shall consider shortly. In the meantime, Albert's result would not apply to testing whether one drug is better than another, i.e., $H_1 : p_1 > p_2$ vs. $H_2 : p_1 \leq p_2$, unless one assumed that the drugs have substantially different effects.

Those problems where the cost per observation depends on the experiment are easily handled by the theory. Here one works with information per unit cost and there is no difficulty in extending the results. Bessler, Chernoff and Marshall [B11] solved the game problem derived from an accelerated life testing problem where the lifetime of an item being stressed by an amount x has an exponential distribution with failure rate $\theta_1 x + \theta_2 x^2$, for $0 \leq x \leq x^*$. It is desired to test whether $\theta_1 x_0 + \theta_2 x_0^2 \leq K$ or $\theta_1 x_0 + \theta_2 x_0^2 > K$, where x_0 is a standard stress. Part of the advantage of using $x \geqq x_0$ (*acceleration*) derives from the fact that the expected time duration to failure is $(\theta_1 x + \theta_2 x^2)^{-1}$. Since an item which failed can promptly be replaced with a new one, and the expense of the experiment is assumed to be proportional to the time spent, it pays to use $x > x_0$. Indeed the optimal design corresponding to a given θ involves at most two stress levels and if two stress levels are used, one is x^*.

Another approach to the sequential design problem was developed by Box and Hill [B4]. To discriminate between two distinct regression models, Box and Hill applied the following method of selecting the next experiment. Select e_{n+1} to maximize

$$(15.1) \qquad \sum_{j>k} \xi_{nj}\xi_{nk}[I(\theta_j, \theta_k, e) + I(\theta_k, \theta_j, e)],$$

where ξ_{nj} represents the posterior probability of θ_j after n observations. This method, which was suggested by a bound on the expected change in entropy resulting from experiment e, is *not* asymptotically optimal except for special circumstances. However, when Meeter, Pirie and Blot [M2] compared the Box-Hill (B-H) method with that proposed by Bessler for two examples, the Monte-Carlo simulations favored the B-H method which seemed to do as well or better.

A study of these simulations suggests that Procedure A can be quite ineffectual in the early stages when the estimates $\hat{\theta}_n$ are poor and tend to trap the experimenter in relatively long series of uninformative experiments which do little to improve the estimate $\hat{\theta}_n$ so that more informative experiments can be performed. Thus, unless one contemplates a very long sequence of experiments with very low error probabilities, the asymptotic efficiency of A may lack sufficient relevance. What is called for is a simple rule which is asymptotically efficient and also effective in the early stages when $\hat{\theta}_n$ is not precise. The use of the posterior probabilities in selecting e_n may add somewhat to the difficulty in implementing a procedure but seems quite sensible in attacking this problem.

Blot and Meeter [B5] have proposed a procedure (B) in which the experimentation rule consists of selecting e_{n+1} to maximize

$$(15.2) \qquad \sum_{\varphi \in a(\hat{\theta}_n)} \xi_{n\varphi}I(\hat{\theta}_n, \varphi, e),$$

and subsequently Chernoff [C17] proposed (M) to select e_{n+1} to minimize

$$\sum_{\theta \in \Theta} \xi_{n\theta}\left[\sum_{\varphi \in a(\theta)} \xi_{n\varphi}I(\theta, \varphi, e) \Big/ \sum_{\varphi \in a(\theta)} \xi_{n\varphi}\right]$$

as a design which would ordinarily lead to asymptotically optimal results. Blot found counterexamples to that claim. Blot and Meeter have obtained conditions under which the procedure (B) and the B-H method are asymptotically optimal for the k-action k state of nature problem. But these are basically rather special conditions on I and as such fail to confront the main issue of how to deal effectively with both the asymptotic and relatively short term effects. Whittle [W4] attempted to face this problem but his approach leads to a solution which is related to the backward induction of dynamic programming and is difficult to implement.

The problem is still open and important.

16. The problem of adjacent hypotheses. Another serious question was that raised by the failure of Procedure A when the sets Θ_i in which action i is preferred, are not separated. Here the major difficulty does not seem to be one of experimental choice but rather one of selecting an appropriate stopping rule.

Consider the case where there is no choice of experiment and the problem is to test $H_1 : \mu \geq 0$ against $H_2 : \mu < 0$, where the observations are independent with distribution $\mathcal{N}(\mu, \sigma^2)$ with σ^2 known. The Wald recommendation would be to select $\mu_1 > 0 > \mu_2$, where (μ_2, μ_1) represents an indifference zone in which it is not very important what decision is made. Then test the simple hypothesis $H_1^* : \mu = \mu_1$ vs. $H_2^* : \mu = \mu_2$, and use this test to decide between H_1 and H_2.

In the symmetric case where $\mu_2 = -\mu_1$, the SPRT for H_1^* vs. H_2^* is of the form:

Stop sampling and accept H_1 if $\sum\limits_{i=1}^{n} X_i \geq a$.

Stop sampling and accept H_2 if $\sum\limits_{i=1}^{n} X_i \leq b$.

Continue sampling otherwise.

For this normal distribution problem, the form of the test does not depend on the choice of μ_1. To evaluate its operating characteristics, i.e., error probability and expected sample size as a function of μ, we may apply the results of § 11.

In this special case where the precise values of μ_1 and $\mu_2 = -\mu_1$ are not important we select them as $\pm 1/2$ for convenience; then

$$Y = \log[f(X|\mu_1)/f(X|\mu_2)] = \sigma^{-2} X$$

and $A \doteq e^{a/\sigma^2}, B = e^{b/\sigma^2}, h(\mu) = -2\mu$, and we have

$$\alpha(\mu) = P_\mu(\text{reject } H_1) \approx \frac{1 - e^{-2\mu a/\sigma^2}}{e^{-2\mu b/\sigma^2} - e^{-2\mu a/\sigma^2}}, \qquad \mu \neq 0,$$

$$E_\mu(N) \approx \frac{(1 - e^{-2\mu a/\sigma^2})b + (e^{-2\mu b/\sigma^2} - 1)a}{\mu(e^{-2\mu b/\sigma^2} - e^{-2\mu a/\sigma^2})}, \qquad \mu \neq 0.$$

Technically, to handle the special case $\mu = 0$, or in general the case where $h(\theta) = 0$ because $E_\theta(Y) = 0$, Wald introduced an extension of the Wald lemma:

$$E_\theta[S_N^2] = E_\theta(N)E_\theta(Y^2) \quad \text{if} \quad E_\theta(Y) = 0,$$

which may be derived from the fact that $S_n^2 - nE_\theta(Y^2)$ is a martingale. In our special case a limiting argument as $\mu \to 0$ can also be used to derive the correct result. We obtain

$$\alpha(0) \approx a/(a - b), \quad E_0(N) \approx -ab\sigma^{-2}.$$

Now consider the symmetric case where $a = -b = -\sigma^2 \log c$. Then

$$\varepsilon(1/2) = \alpha(1/2) = \varepsilon(-1/2) = 1 - \alpha(-1/2) \sim c,$$

$$E_{1/2}(N) = E_{-1/2}(N) \approx 2\sigma^2(-\log c)$$

while

$$E_0(N) \sim \sigma^2(-\log c)^2, \quad \alpha(0) \sim 1/2.$$

In this example the cost of sampling in the least important case, when $\mu = 0$, *is larger by an order of magnitude* than the risk for $\mu = \pm 1/2$. Now it is perfectly reasonable that as $\mu \to 0$, the inability to distinguish whether $\mu > 0$ or $\mu < 0$ should force the cost of sampling up. But it is disturbing that the cost should be increased so much.[3] One may ask whether an improved stopping rule can keep costs down to the order of $O(-c \log c)$.

Kiefer and Weiss [K5] attacked this problem by considering the three state of nature problem where the statistician minimizes $E_{\theta_0}(N)$ subject to specified error probabilities under θ_1 and θ_2. Here θ_0 was regarded as a point of indifference as to the terminal action. The criterion neglected explicit consideration of $E_{\theta_1}(N)$ and $E_{\theta_2}(N)$. Considerably more insight was obtained by Schwarz [S7] who phrased the three state problem from a decision point of view.

The three possible states of nature are $\theta_0, \theta_1, \theta_2$. The losses are given by Table 3.

TABLE 3

	a_1	a_2
θ_0	0	0
θ_1	0	$r(\theta_1)$
θ_2	$r(\theta_2)$	0

Thus $R(\theta_0) = cE_{\theta_0}(N)$, $R(\theta_1) = cE_{\theta_1}(N) + r(\theta_1)\varepsilon(\theta_1)$ and $R(\theta_2) = cE_{\theta_2}(N) + r(\theta_2)\varepsilon(\theta_2)$, where $\varepsilon(\theta_1) = P_{\theta_1}(\text{reject } H_1) = \alpha(\theta_1)$, and $\varepsilon(\theta_2) = P_{\theta_2}(\text{accept } H_1) = 1 - \alpha(\theta_2)$. In this formulation θ_0 is a point of indifference, but a good procedure must consider the expected sample size as well as the error probabilities when $\theta = \theta_1$ or θ_2.

Schwarz derived bounds for Bayes procedures for this problem when $f(x|\theta)$ belongs to an exponential family of distributions, i.e., a family for which

$$f(x|\theta) = \exp[-a(\theta) + b(x) + c(x)d(\theta)]$$

and for which many desirable properties hold and which contains many of the well-known special distributions. In particular, he demonstrated that for the problem of testing $H_1 : \mu = \mu_1 > 0$ vs. $H_2 : \mu = \mu_2 < 0$, where $\mu = 0$ is possible and the data are independent $\mathcal{N}(\mu, \sigma^2)$ with known σ^2, the Bayes procedure can be approximated by a polygonal figure in the (n, S_n)-space, $S_n = \sum_{i=1}^{n} X_i$.

---

[3] The factor $(\log c)$ is more disturbing theoretically than practically because of the slow rate at which $\log c$ increases.

Rather than use the refined approximations of Schwarz, let us continue with the moderately crude approximations used previously in our theory of asymptotically optimal sequential design. This cruder approach does not require $f(x|\theta)$ to belong to an exponential family.

The posterior probabilities of θ_0, θ_1, θ_2 are proportional to the priors multiplied by the likelihoods. Thus we have

$$\xi_{0n}:\xi_{1n}:\xi_{2n} = \xi_0 L_n(Z_n|\theta_0):\xi_1 L_n(Z_n|\theta_1):\xi_2 L_n(Z_n|\theta_2),$$

where

$$L_n(Z_n|\theta) = \prod_{i=1}^{n} f(X_i|\theta)$$

is the likelihood of θ based on n observations. The posterior risks associated with stopping and deciding are proportional to

$$r(\theta_2)\xi_2 L_n(Z_n|\theta_2) \quad \text{for action 1}$$

and

$$r(\theta_1)\xi_1 L_n(Z_n|\theta_1) \quad \text{for action 2.}$$

If θ is the true state of nature and $f(x|\theta) \neq f(x|\varphi)$, then

$$\frac{L_n(Z_n|\theta)}{L_n(Z_n|\varphi)} = e^{S_n(\theta,\varphi)} \approx e^{nI(\theta,\varphi)},$$

and hence $\xi_n(\theta)/\xi_n(\varphi)$ grows exponentially fast. From our crude point of view, the priors and $r(\theta_i)$ are relatively unimportant and the risk associated with stopping will be of the order of magnitude of c if we stop and decide for action 1 when either

$$\frac{L_n(Z_n|\theta_1)}{L_n(Z_n|\theta_2)} = e^{S_n(\theta_1,\theta_2)} \geq c^{-1}$$

or

$$\frac{L_n(Z_n|\theta_0)}{L_n(Z_n|\theta_2)} = e^{S_n(\theta_0,\theta_2)} \geq c^{-1} \quad \text{and} \quad L_n(Z_n|\theta_1) > L_n(Z_n|\theta_2).$$

A similar statement applies for stopping and deciding for action 2. As in our former theory, the attractiveness of a stopping risk of the order of c comes from the consideration of minimizing an expression of the form

$$cn + re^{-nI}.$$

Abbreviating and recapitulating, we represent the likelihoods by L_{0n}, L_{1n}, L_{2n}. We stop and accept $H_1 : \theta = \theta_1$ if

(16.1) $$\max(L_{0n}, L_{1n})/L_{2n} \geq c^{-1} \quad \text{and} \quad L_{1n} \geq L_{2n}.$$

We stop and accept $H_2 : \theta = \theta_2$ if

(16.2) $\max(L_{0n}, L_{2n})/L_{1n} \geq c^{-1}$ and $L_{2n} > L_{1n}$.

Otherwise we continue sampling.

If θ_1 is the true state, then

$$L_{1n}/L_{2n} \approx e^{nI(\theta_1, \theta_2)}, \quad L_{1n}/L_{0n} \approx e^{nI(\theta_1, \theta_0)},$$

and one expects

$$N \approx \frac{-\log c}{I(\theta_1, \theta_2)}.$$

If θ_2 is the true state one expects, by symmetry,

$$N \approx \frac{-\log c}{I(\theta_2, \theta_1)}.$$

If θ_0 is the true state, then

$$L_{0n}/L_{1n} \approx e^{nI(\theta_0, \theta_1)}, \quad L_{0n}/L_{2n} \approx e^{nI(\theta_0, \theta_2)}$$

and one expects

$$N \approx \frac{-\log c}{I(\theta_0)},$$

where

(16.3) $I(\theta_0) = \max[I(\theta_0, \theta_1), I(\theta_0, \theta_2)].$

Let us examine what this suggested procedure becomes in the following normal distribution example.

Example 16.1. $X \sim \mathcal{N}(\theta, 1)$, $\theta_0 = 0$, $\theta_1 = 1/2$, $\theta_2 = -1/2$,

$$S_n(\theta, \varphi) = (\theta - \varphi)\left[S_n - n\left(\frac{\theta + \varphi}{2}\right)\right], \quad S_n = \sum_{i=1}^{n} X_i.$$

Then

$$\max(L_{1n}, L_{0n})/L_{2n} \geq c^{-1}$$

when

$$\max[S_n(\theta_1, \theta_2), S_n(\theta_0, \theta_2)] \geq -\log c$$

or

(16.4) $\max\left[S_n, \frac{1}{2}\left(S_n + \frac{n}{4}\right)\right] \geq -\log c.$

Comparing conditions for $\max(L_{2n}, L_{0n})/L_{1n} \geq c^{-1}$ and for $L_{1n} > L_{2n}$, we obtain the pentagon of Fig. 10 where $k = -\log c$.

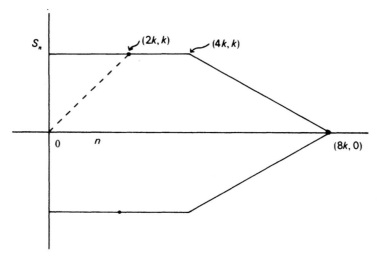

FIG. 10

This pentagon has three special points $(2k, k), (8k, 0)$ and $(2k, -k)$. If $\theta = \theta_1 = 1/2$, S_n tends to be roughly equal to $n/2$ and the path traced by (n, S_n) tends to be close to the straight line from the origin to $(2k, k)$. Similarly, if $\theta = \theta_0 = 0$, the path tends to be close to the horizontal line from $(0, 0)$ to $(8k, 0)$. This suggests that except for the three special points, there may be scope for modifying the figure without seriously affecting performance for $\theta = \theta_0, \theta_1, \theta_2$. For example, we might be able to replace the line from $(0, k)$ to $(2k, k)$ by something lower without hurting performance. Why should we do this if it does not help? If we consider extending the problem to the case where θ might take on other values, then modifying the boundary to be appropriate for these other values too may be desirable.

Thus the Schwarz solution seems to carry some flexibility which leaves the possibility of further extension to the case of more than three possible states of nature. Schwarz noted that when dealing with the two-action problem with five possible θ, he was led to a seven-sided polygon. Some consideration of that solution suggested to him that his reasoning extended to the following model where there is a true indifference zone.

Let Θ be the union of three disjoint sets Θ_0, Θ_1 and Θ_2, where action i is preferred for $\theta \in \Theta_i$, $i = 1, 2$. On Θ_0, either action may be taken. Thus we have

$$r(\theta) = l(\theta, a_2) > 0, \quad l(\theta, a_1) = 0 \quad \text{for} \quad \theta \in \Theta_1,$$

$$r(\theta) = l(\theta, a_1) > 0, \quad l(\theta, a_2) = 0 \quad \text{for} \quad \theta \in \Theta_2,$$

$$r(\theta) = l(\theta, a_1) = l(\theta, a_2) = 0 \quad \text{for} \quad \theta \in \Theta_0.$$

Let

$$L_{in} = \sup_{\theta \in \Theta_i} \prod_{j=1}^{n} f(X_j | \theta) \qquad i = 0, 1, 2,$$

be the maximum likelihood for $\theta \in \Theta_i$. We propose:

Stop and accept $H_1 : \theta \in \Theta_1$ *if* $\max(L_{1n}, L_{0n})/L_{2n} \geqq c^{-1}$, $L_{1n} \geqq L_{2n}$,

Stop and accept $H_2 : \theta \in \Theta_2$ *if* $\max(L_{2n}, L_{0n})/L_{1n} \geqq c^{-1}$, $L_{2n} > L_{1n}$,

Otherwise continue sampling.

We return to extend the normal example.

Example 16.2. Test $H_1 : \theta \geqq 1/2$ vs. $H_2 : \theta \leqq -1/2$ when it is also possible that $-1/2 < \theta < 1/2$.

Then we use the fact that

$$L_{in}/L_{jn} = \exp[S_{ijn}],$$

where

$$S_{ijn} = \sup_{\theta \in \Theta_i} \inf_{\varphi \in \Theta_j} S_n(\theta, \varphi),$$

$$S_n(\theta, \varphi) = (\theta - \varphi)\left[S_n - n\left(\frac{\theta + \varphi}{2}\right)\right].$$

If $\bar{X}_n \geqq 1/2$, $L_{1n} \geqq L_{0n} > L_{2n}$ and

$$S_{12n} = \left(S_n + \frac{n}{2}\right)^2 \bigg/ 2n.$$

If $0 \leqq \bar{X}_n < 1/2$, $L_{0n} > L_{1n} \geqq L_{2n}$ and

$$S_{02n} = \left(S_n + \frac{n}{2}\right)^2 \bigg/ 2n.$$

The case of $\dot{X}_n < 0$ can be treated symmetrically. Compare the stopping boundary $(\pm S_n + n/2)^2/2n = k$ represented in Fig. 11 with the pentagon of the preceding

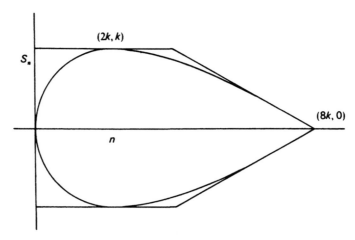

FIG. 11

example. Note that the boundary consists of two parabolas with tilted axes, which have infinite slope at $(0, 0)$, zero slope at $(2k, k)$ and slope $\pm \frac{1}{4}$ at $(8k, 0)$. Thus the new stopping region is within the pentagon but goes through the special points so as to be tangent to the pentagon.

The considerations which led to this approach suggest that the error probabilities will be of order c and that the expected sample size will satisfy

$$E_\theta(N) \approx \frac{-\log c}{I(\theta)},$$

where

$$I(\theta) = \begin{cases} \inf_{\varphi \in \Theta_2} I(\theta, \varphi) & \text{for } \theta \in \Theta_1, \\ \inf_{\varphi \in \Theta_1} I(\theta, \varphi) & \text{for } \theta \in \Theta_2, \\ \max\left[\inf_{\varphi \in \Theta_1} I(\theta, \varphi), \inf_{\varphi \in \Theta_2} I(\theta, \varphi)\right] & \text{for } \theta \in \Theta_0, \end{cases}$$

thereby keeping our risks

$$R(\theta) \approx \frac{-c \log c}{I(\theta)}$$

in this formulation where the *indifference regions separate the alternative hypotheses*.

Returning to our example it is easy to see that $I(\theta, \varphi) = (\theta - \varphi)^2/2$, and

(16.5) $$I(\theta) = (|\theta| + \tfrac{1}{2})^2/2 \geqq 1/8.$$

Let us review the status of our problem. We developed a method of sequential design of experiments which has two serious shortcomings. One is that though its efficacy is great when θ is well known, there is danger in the early stages of sampling that it will lead to a poor choice of experiments, and so far there is no general feasible approach to deal neatly with that problem. The second difficulty is that asymptotic optimality breaks down by an order of magnitude near the common boundary if the two hypotheses are not separated.

To treat this latter problem Schwarz attacked the case of three possible states of nature, for one of which either decision was equally good. His results extended naturally to the case of three sets of θ: Θ_0, Θ_1, and Θ_2, where Θ_1 and Θ_2 are separated by an indifference zone Θ_0. We have presented a crude but general version of his results here. However, these results are of dubious significance. For the problem of testing whether the mean θ of a normal distribution is positive or negative, the natural loss function has a loss

$$r(\theta) \approx k|\theta| \quad \text{for } |\theta| \text{ close to } 0,$$

and though this is small for θ small, it is not negligible. Theoretically, the indifference zone is not a satisfactory concept and the Schwarz strategy led to an elegant solution whose relevance is dubious and must be examined closely in special cases.

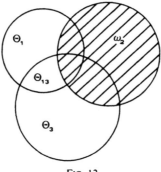

FIG. 12

We propose to return to deal with the question of "no indifference zone" shortly. First, let us summarize the Kiefer–Sacks result [K3] which combined the results of Chernoff, Albert, Bessler and Schwarz.

For illustration it is convenient to think in terms of the three-action problem where the parameter space Θ is decomposed into the pairwise disjoint sets Θ_i where action a_i is preferred, Θ_{ij} where actions a_i and a_j are equally good and better than the remaining action and the Θ_{123} where all three actions are equally good.

$$\Theta = \Theta_1 \cup \Theta_2 \cup \Theta_3 \cup \Theta_{12} \cup \Theta_{13} \cup \Theta_{23} \cup \Theta_{123}.$$

The sets ω_i are formed by combining all those subsets which contain i as a subscript (see Fig. 12). For example,

$$\omega_2 = \Theta_2 \cup \Theta_{12} \cup \Theta_{23} \cup \Theta_{123}$$

is the set on which a_2 is optimal. From a Bayesian or asymptotic point of view it is reasonable to start with some nondegenerate prior probabilities and to stop when the posterior risk for one of the actions is reduced below some number of the order of magnitude of c and to select the action for which the corresponding posterior risk is a minimum. However, the risks approach zero in a manner determined by the rates at which the likelihood ratios approach zero, and it is these rates which help determine the appropriate information numbers and experimental choice.

Let $l(\theta, a) \geq 0$ be a loss function defined for $\theta \in \Theta$ and $a \in (a_1, a_2, \cdots, a_k)$ such that for each θ there is at least one best action a for which $l(\theta, a) = 0$, and for each action a there is at least one θ for which $l(\theta, a) = 0$. The latter condition states that there are no *hedging* actions for such actions would have no impact in an asymptotic theory where the cost c per observation approaches zero.

Let $G_0 = \{1, 2, \cdots, k\}$ and we shall use G to denote a *nonnull* subset of G_0. Let

(16.6) $\Theta_G = \{\theta : l(\theta, a_i) = 0 \Leftrightarrow i \in G\},$ $\varnothing \neq G \subset G_0,$

and

(16.7) $\omega_i = \bigcup_{i \in G} \Theta_G = \{\theta : l(\theta, a_i) = 0\},$ $1 \leq i \leq k,$

represent the set on which any action a_i for $i \in G$ is optimal, and the set on which $l(\theta, a_i) = 0$, respectively.

Let the maximum likelihoods on ω_i and ω_i^c be given by

$$(16.8) \qquad P_{in} = \sup_{\theta \in \omega_i} \prod_{i=1}^{n} f(X_j | \theta, e_j),$$

$$(16.9) \qquad \bar{P}_{in} = \sup_{\theta \notin \omega_i} \prod_{j=1}^{n} f(X_j | \theta, e_j).$$

Asymptotically the rule of stopping and choosing a terminal action in terms of the posterior risk relates to:

$$\text{Stop as soon as } P_{in}/\bar{P}_{in} > c^{-1} \text{ for some } i$$

$$\text{and select } a_i \text{ for which } P_{in}/\bar{P}_{in} \text{ is a maximum.}$$

If we concentrate for the moment on the case where there is no choice of experiment, we are led to

$$I(\theta) = \max_{i \in G} \inf_{\varphi \notin \omega_i} I(\theta, \varphi) \quad \text{for} \quad \theta \in \Theta_G.$$

The reader may find it helpful to relate this to our illustration keeping in mind that $\inf_{\varphi \in A} I(\theta, \varphi)$ may be regarded as a *distance* from θ to A, and is sometimes written $I(\theta, A)$. Thus $\inf_{\varphi \notin \omega_i} I(\theta, \varphi) = I(\theta, \omega_i^c)$, the distance from θ to the set where a_i is not optimal. If $\theta \in \Theta_{12}$, then $I(\theta)$ is the distance from θ to the set where a_1 is not optimal or to the set where a_2 is not optimal, whichever is larger.

The natural choice of experimentation is quickly seen to be:

$$(16.10) \qquad \begin{array}{c} \text{If } \hat{\theta}_n \in \Theta_G \text{ select } e^{(n+1)} \in \mathscr{E}^* \text{ to maximize} \\[6pt] \max_{i \in G} \inf_{\varphi \notin \omega_i} I(\hat{\theta}_n, \varphi, e). \end{array}$$

The resulting procedure will have error probabilities of order c and expected sample sizes

$$E_\theta(N) \approx \frac{-\log c}{I(\theta)},$$

where

$$(16.11) \qquad I(\theta) = \sup_{e \in \mathscr{E}^*} \max_{i \in G} \inf_{\varphi \notin \omega_i} I(\theta, \varphi, e) \quad \text{for} \quad \theta \in \Theta_G$$

gives asymptotically optimal risks

$$(16.12) \qquad R(\theta) \approx \frac{-c \log c}{I(\theta)}.$$

Kiefer and Sacks also observed that from the point of view of asymptotic optimality, one may proceed in two stages. In the first stage select $n_1 = \sqrt{-\log c}$

observations on an experiment (possibly randomized) which guarantees consistency, i.e., $\hat{\theta}_n \to \theta$. Since n_1 is large, $\hat{\theta}_{n_1}$ will be close to the true θ and the remaining experiments can be selected from the $e \in \mathscr{E}^*$ whose maximum yields $I(\hat{\theta}_{n_1})$. While this two-stage approach theoretically eliminates the need for major recomputation after each observation, it tends to remove the ability to make valuable reconsiderations as evidence appears. In this way it simply aggravates the major defect of Procedure A that it is a very large sample theory inappropriate for moderate sample sizes and not especially instructive for the early stages of experimentation.

One should note that the main function of randomization in this sequential design theory is merely to see that various experiments are performed in certain proportions. Thus, in the two-stage approach and even using Procedure A where the appropriate proportions of $e_n \in \mathscr{E}^*$ change slowly with n, it is possible to replace randomized experiments by deterministic sequences in which the various elementary experiments of $\mathscr{E}$ are performed in appropriate proportions.

17. Testing for the sign of a normal mean: no indifference zone. We return to the problem of dealing with composite hypotheses without indifference zones. Let us concentrate our attention on the special problem of testing whether the mean of a normal distribution with known variance σ^2 is positive or negative. We shall take the cost of a wrong decision to be

$$(17.1) \qquad\qquad r(\mu) = k|\mu|$$

and the cost per observation is c.

The advantage of this example lies partly in the fact that the normal distribution is generally an easy distribution to deal with. More important, however, is that in large sample theory many important statistics are asymptotically normally distributed. Thus the behavior of asymptotically optimal strategies tends to relate to the solution of corresponding problems for normally distributed random variables. Hopefully the solution of the normal distribution problem can be plugged in, with minor modification, to a more general problem and yield asymptotically optimal procedures for the more general problems.

In the nonasymptotic case, the concept of solution is not well specified unless we assume a prior distribution. Generally, the Bayes solution corresponding to a nondegenerate prior tends to converge in the asymptotic case to an asymptotically optimal solution which is independent of the particular prior distribution.

Thus we shall impose a convenient prior distribution on the unknown mean μ, that is, the normal distribution $\mathcal{N}(\mu_0, \sigma_0^2)$.

We summarize the problem.

PROBLEM 17.1. $X_1, X_2, \cdots$ are independent $\mathcal{N}(\mu, \sigma^2)$ variables, σ^2 known. It is desired to test $H_1 : \mu \geqq 0$ vs. $H_2 : \mu < 0$, where the cost of a wrong decision is $r(\mu) = k|\mu|$ and the cost of observing n X's is cn. The parameter μ has the $\mathcal{N}(\mu_0, \sigma_0^2)$ distribution. What is the Bayes sequential strategy?

17.1. Posterior distributions and risks. As a first step we characterize the posterior probability distribution and the posterior risks associated with stopping. First we generalize slightly to the case where the X_i are independent $\mathcal{N}(\mu, \sigma_i^2)$ variables.

LEMMA 17.1.1. *The posterior distribution of μ given $X_1, X_2, \cdots, X_n$ is*

(17.1.1)
$$\mathcal{L}(\mu | X_1, X_2, \cdots, X_n) = \mathcal{N}(Y_n, s_n),$$

where

(17.1.2)
$$Y_n = \frac{\mu_0 \sigma_0^{-2} + X_1 \sigma_1^{-2} + \cdots + X_n \sigma_n^{-2}}{\sigma_0^{-2} + \sigma_1^{-2} + \cdots + \sigma_n^{-2}}$$

and

(17.1.3)
$$s_n^{-1} = (\sigma_0^{-2} + \sigma_1^{-2} + \cdots + \sigma_n^{-2}).$$

We digress briefly to derive this result which expresses the conditional mean of μ given the data as a weighted average of μ_0 and the observations weighted inversely as the variances. The mean of the prior is treated as an observation. Once we demonstrate Lemma 17.1.1 for $n = 1$, the result follows by regarding the posterior distribution based on n observations as the result of the nth observation converting the posterior based on $n - 1$ observations. To demonstrate Lemma 17.1.1 for $n = 1$ there are two approaches. One is to regard μ and X_1 as variables with a joint normal distribution where

$$\mu = \mu_0 + \sigma_0 W_1,$$

$$X_1 = \mu + \sigma_1 W_2 = \mu_0 + \sigma_0 W_1 + \sigma_1 W_2$$

with W_1 and W_2 independent $\mathcal{N}(0, 1)$ variables. Then the joint distribution of μ and X_1 is easily computed to be

$$\mathcal{L}\begin{pmatrix} \mu \\ X_1 \end{pmatrix} = \mathcal{N}\left(\begin{pmatrix} \mu_0 \\ \mu_0 \end{pmatrix}, \begin{pmatrix} \Sigma_{\mu\mu} & \Sigma_{\mu X_1} \\ \Sigma_{X_1 \mu} & \Sigma_{X_1 X_1} \end{pmatrix} \right) = \mathcal{N}\left(\begin{pmatrix} \mu_0 \\ \mu_0 \end{pmatrix}, \begin{pmatrix} \sigma_0^2 & \sigma_0^2 \\ \sigma_0^2 & \sigma_0^2 + \sigma_1^2 \end{pmatrix} \right).$$

The posterior distribution is

$$\mathcal{L}(\mu | X_1) = \mathcal{N}[\mu_0 + \Sigma_{\mu X_1} \Sigma_{X_1 X_1}^{-1}(X_1 - \mu_0), \Sigma_{\mu\mu} - \Sigma_{\mu X_1} \Sigma_{X_1 X_1}^{-1} \Sigma_{X_1 \mu}]$$

by application of a standard formula for conditional distribution using the multivariate normal model. With a certain amount of algebra the lemma follows.

Another approach which comes out neatly is to recall that the posterior is proportional to the prior times the likelihood:

$$g(\mu | X_1) \propto \frac{1}{\sqrt{2\pi\sigma_0^2}} e^{-(\mu - \mu_0)^2 / 2\sigma_0^2} \cdot \frac{1}{\sqrt{2\pi\sigma^2}} e^{-(X_1 - \mu)^2 / 2\sigma_1^2}.$$

Collecting terms with μ in the exponent we have

$$-\frac{1}{2}\left[\mu^2 \left(\frac{1}{\sigma_0^2} + \frac{1}{\sigma_1^2} \right) - 2\mu \left(\frac{\mu_0}{\sigma_0^2} + \frac{X_1}{\sigma_1^2} \right) + \cdots \right]$$

and completing the square this gives

$$-\frac{1}{2}\left[\frac{1}{(\sigma_0^{-2} + \sigma_1^{-2})^{-1}}\right]\left[\mu - \frac{\mu_0\sigma_0^{-2} + X_1\sigma_1^{-2}}{\sigma_0^{-2} + \sigma_1^{-2}}\right]^2 + \cdots$$

which uniquely determines the conditional distribution of μ given X_1 to be the specified normal distribution.

Let us return to Y_n, the conditional mean of μ given $X_1, X_2, \cdots, X_n$. How does Y_n change as additional information comes in? It is easy to anticipate that $E(Y_n - Y_m) = 0$ for $n > m$. What is more surprising is that $Y_n - Y_m$ is *normal and independent of Y_m, and has mean 0 and variance $s_m - s_n$.* We have the following lemma.

LEMMA 17.1.2.

$$\mathscr{L}(Y_n - Y_m | Y_m) = \mathscr{N}(0, s_m - s_n), \qquad\qquad n \geq m \geq 0.$$

Proof. From the definition of $Y_n = E(\mu | X_1, \cdots, X_n)$ and the fact that $E[E(X|Y, Z)|Y] = E(X|Y)$, it follows that

$$Y_m = E(Y_n | X_1, \cdots, X_m)$$

is the regression of Y_n on $(X_1, \cdots, X_m)$. Thus

$$Y_n = Y_m + u,$$

where the residual u has mean 0 and is uncorrelated with Y_m and hence,

$$\sigma_{Y_n}^2 = \sigma_{Y_m}^2 + \sigma_u^2.$$

Since Y_n and Y_m are linear functions of μ_0 and the X_i, Y_m and $u = Y_n - Y_m$ have a joint normal distribution and then zero correlation implies independence. It remains only to show that $\sigma_u^2 = \sigma_{Y_n}^2 - \sigma_{Y_m}^2 = s_m - s_n$. But Y_n is the regression of μ, and $\mu = Y_n + u_n$, where $\sigma_\mu^2 = \sigma_0^2$ and $\sigma_{u_n}^2 = s_n$, and hence, $\sigma_{Y_n}^2 = \sigma_0^2 - s_n$. The desired equality follows for σ_u^2.

These two lemmas indicate that as evidence accumulates, the posterior mean of μ behaves like a Gaussian stochastic process with independent increments, starting from $Y_0 = \mu_0$.

It is interesting that the numerator of Y_n,

$$\mu_0\sigma_0^{-2} + X_1\sigma_1^{-2} + \cdots + X_n\sigma_n^{-2}$$

is a process of independent increments for fixed μ. If μ is regarded as a random variable, the numerator loses that property but Y_n then achieves this property.

The expression

$$s_n^{-1} = \sigma_0^{-2} + \sigma_1^{-2} + \cdots + \sigma_n^{-2}$$

is called the *precision* of the estimate Y_n. As n increases s_n decreases. In the special case of main interest where $\sigma_1^2 = \sigma_2^2 = \cdots = \sigma_n^2 = \sigma^2$ and the X_i are independent (for fixed μ) with common distribution,

$$(17.1.2')\qquad\qquad Y_n = \frac{\mu_0\sigma_0^{-2} + (X_1 + \cdots + X_n)\sigma^{-2}}{\sigma_0^{-2} + n\sigma^{-2}},$$

and the precision

$$(17.1.3') \qquad\qquad s_n^{-1} = \sigma_0^{-2} + n\sigma^{-2}$$

increases linearly in n.

Now let us compute the posterior risk of stopping and accepting the hypothesis H_1. Since the loss is $k|\mu|$ when μ is less than 0 and 0 otherwise, we have a risk of

$$\int_{-\infty}^{0} k|\mu| n(\mu; y, s)\, d\mu,$$

where $n(\mu; y, s)$ represents the density of a random variable with distribution $\mathcal{N}(y, s)$. The quantity is readily computed and found to be $ks^{1/2}\psi^+(ys^{-1/2})$, where $\psi^+(u) = \varphi(u) - u[1 - \Phi(u)]$, with φ the standard normal density and Φ the standard normal c.d.f. Similarly, the posterior risk of stopping and rejecting H_1 is $ks^{1/2}\psi^-(ys^{-1/2})$, where $\psi^-(u) = \varphi(u) + u\Phi(u)$. It is no surprise that $\psi^+(u) < \psi^-(u)$ when $u > 0$. The functions $\psi^+(u)$ and $\psi^-(u)$ have some interesting properties: (i) $\psi^+(u) = \psi^-(u) - u$, (ii) $\psi^+(u) = \psi^-(-u)$, (iii) $\psi^+(u) \sim u^{-2}\varphi(u)$ $\cdot \{1 - 3u^{-2} + 15u^{-4} - \cdots\}$ as $u \to \infty$.

Returning to our sequential analysis problem, we have the risk associated with stopping and deciding for or against H_1 plus the cost of sampling cn $= c(s_n^{-1} - \sigma_0^{-2})\sigma^2$. Thus we have the following lemma.

LEMMA 17.1.3. *The posterior risk associated with stopping at the nth observation in the sequential analysis problem is $d(Y_n, s_n)$, where*

$$(17.1.4) \qquad d(y, s) = ks^{1/2}\psi(ys^{-1/2}) + c\sigma^2 s^{-1} - c\sigma_0^{-2}\sigma^2$$

and

$$(17.1.5) \qquad \psi(u) = \begin{cases} \varphi(u) - u[1 - \Phi(u)], & u \geq 0, \\ \varphi(u) + u\Phi(u), & u \leq 0. \end{cases}$$

With this lemma the problem of finding the Bayes procedure for our sequential analysis problem has been reduced to a *stopping problem* of the type described in the next subsection.

17.2. Stopping problems. Let $(Y_n, s_n, n \in G)$ be a Gaussian process of independent increments starting from (Y_{n_0}, s_{n_0}), $n_0 \in G$, with

$$(17.2.1) \qquad \mathcal{L}(Y_n - Y_m | Y_m) = \mathcal{N}(0, s_m - s_n), \qquad n \geq m,$$

and let the cost associated with stopping at (Y_n, s_n) be given by $d(Y_n, s_n)$. Find a stopping rule (a random variable N taking on values in G such that $N < \infty$ with probability one and $\{N = n\} \in \mathcal{B}\{Y_i : i \in G, n_0 \leq i \leq n\}$) so as to minimize

$$(17.2.2) \qquad E[d(Y_N, s_N)].$$

In addition to the sequential analysis problem of § 17.1 we could consider another example of a stopping problem.

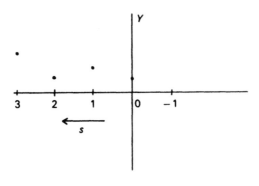

Example 17.2.1. Let Y_n be a process starting at y_0 for $n = n_0$, n_0 a negative integer. As n increases let $Y_{n+1} = Y_n + u_n$, where the u_n are independent $\mathcal{N}(0, 1)$ random variables. The player may stop or, at a cost of one, increase n by one. If he continues till $n = 0$, the game ends and he receives 0 if $Y_0 \geq 0$ and Y_0^2 if $Y_0 \leq 0$.

Let $s_n = -n$, $G = \{n : n$ is an integer $\leq 0\}$, $Y_{n_0} = y_0$, and

$$d(y, s) = \begin{cases} -y^2 - s - n_0 & \text{if } s = 0 \text{ and } y \leq 0, \\ -s - n_0 & \text{otherwise,} \end{cases}$$

and this example fits into our framework. The reader may find it convenient to use (y, s) graphs where s decreases as one moves to the right (see Fig. 13).

Note that in this example it seems natural to expect to invest in continuing if Y_n is highly negative and to stop if Y_n is highly positive.

17.3. Continuous time stopping problems. Consider the sequential analysis problem when the cost of sampling is small. Since

$$Y_n = \frac{\mu_0 \sigma_0^{-2} + (X_1 + \cdots + X_n)\sigma^{-2}}{\sigma_0^{-2} + n\sigma^{-2}}$$

the decision to stop or continue after the nth observation depends on $(X_1 + \cdots + X_n)$. (In fact this sum is a *sufficient statistic* for μ, where the term sufficient is a technical term used to indicate that the statistic contains all the relevant information for inference on μ that is present in the data.) Thus we are interested in the process $\sum_{i=1}^n X_i$ which is for fixed μ a Gaussian process of independent increments. When c is small, we expect to observe this sum for many values of n. When suitably reduced in scale, a graph of this process will resemble a continuous function with somewhat erratic behavior. A limiting version of this process is Brownian motion or the Wiener process where the sum $X_1 + \cdots + X_n$ is replaced by the continuous Gaussian process $X(t)$ with independent increments described by

$$E[dX(t)] = \mu \, dt,$$

(17.3.1)

$$\text{var}[dX(t)] = \sigma^2 \, dt.$$

When μ is regarded as a random variable, the limiting form of the (Y_n, s_n) process is the Gaussian process of independent increments (in the $-s$ scale), $Y(s)$, where

(17.3.2)
$$E[dY(s)] = 0,$$
$$\text{var}[dY(s)] = -ds$$

with $Y(s_0) = \mu_0$ at $s_0 = \sigma_0^2$ and

(17.3.3)
$$s^{-1} = \sigma_0^{-2} + t\sigma^{-2}.$$

Note that as t increases from 0 to ∞, s decreases from σ_0^2 to 0. Thus $(-ds)$ may be thought of as positive. Hence a limiting form of our sequential problem is a special case, for $G = (0, \infty)$, of the following stopping problem.

STOPPING PROBLEM. Given a Gaussian process $\{Y(s), s \in G\}$ of independent increments in the $-s$ scale, with $EdY(s) = 0$, $\text{var}[dY(s)] = -ds$, starting at $Y(s_0) = y_0$, find a stopping time S (S is a r.v. on G, where $\{S < s\} \in \mathcal{B}\{Y(s'): s_0 \geq s' \geq s\}$) to minimize

(17.3.4)
$$E[d(Y(S), S)].$$

The continuous time version of the sequential analysis problem is essentially associated with the function $d(y, s)$. We have

$$d(y, s) = ks^{1/2}\psi(ys^{-1/2}) + c\sigma^2 s^{-1} - c\sigma_0^{-2}\sigma^2, \qquad \sigma_0^2 \geq s > 0.$$

Not only is the continuous time problem a limiting form of the discrete time problem, but we may regard the latter as a special variation of the continuous time problem subject to the restriction that stopping may take place only at certain specified values of s, i.e., $s = (\sigma_0^{-2} + t\sigma^{-2})^{-1}$ for integer values of t. In this continuous time framework, Y is regarded as a function of s and the subscript n has been eliminated as an unnecessary parameter which serves only to mark the intervals between possible stopping times.

From the point of view of solving the sequential analysis problem, certain simplifying transformations can be made. First the constant $-c\sigma_0^{-2}\sigma^2$ has no effect on the optimal procedure.

Second, the transformation

(17.3.5)
$$Y^*(s^*) = aY(s),$$
$$s^* = a^2 s$$

converts the $Y(s)$ to the $Y^*(s^*)$ process which is also a Gaussian process of independent increments with $EdY^*(s^*) = 0$, $\text{var}[dY^*(s^*)] = a^2 \text{var}[dY(s)] = -a^2 ds = -ds^*$. Then taking $a = k^{1/3}c^{-1/3}\sigma^{-2/3}$,

(17.3.6)
$$d(y, s) = k^{2/3}c^{1/3}\sigma^{2/3} d^*(y^*, s^*) - c\sigma_0^{-2}\sigma^2,$$

where

$$d^*(y^*, s^*) = s^{*1/2}\psi(y^*s^{*-1/2}) + (s^*)^{-1}.$$

Thus our problem may be normalized by this transformation to that of dealing with the stopping cost d^*. Note that if c is small, s is multiplied by a large factor to get s^*. Thus $s = \sigma_0^2$ corresponds to $s^* = a^2\sigma_0^2$. In our normalized problem the prior distribution has almost negligible precision. When s^* reaches ordinary levels, a change of one in t results in a change of $-(s^*)^2/a^2\sigma^2$ in $s^* = a^2(\sigma_0^{-2} + t\sigma^{-2})^{-1}$. Thus each observation then leads to a *small* change in s^*, a change of order $c^{2/3}$.

We have the following example.

Example 17.3.1 (Normalized continuous time sequential analysis problem).

$$(17.3.7) \qquad d(y, s) = s^{1/2}\psi(ys^{-1/2}) + s^{-1}, \qquad\qquad s > 0.$$

By adding a constant in Example 17.2.1, its continuous time version may be expressed as follows.

Example 17.3.2.

$$(17.3.8) \qquad d(y, s) = \begin{cases} -y^2 - s & \text{if } s = 0 \text{ and } y \leq 0, \\ -s & \text{otherwise for } s \geq 0. \end{cases}$$

17.4. Continuous time problems: relevance of stopping sets. If the original discrete time sequential analysis problem were truncated so that stopping must take place by the n_1st observation for some n_1, the resulting problem could be attacked by the backward induction method of dynamic programming. The same could be said for the original (discrete time) version of Example 17.2.1. This method can be summarized by the equation

$$(17.4.1) \qquad \rho_n(\xi_n) = \inf_{a_n} E\{\rho_{n+1}[\xi_{n+1}(a_n, \xi_n)]\},$$

where $\rho_n(\xi_n)$ is the expected cost of an optimal procedure given the history ξ_n up to stage n, $\xi_{n+1}(a_n, \xi_n)$ describes the possibly random history up to stage $n + 1$ with probability distribution depending on ξ_n and the action a_n taken at stage n.

To illustrate with Example 17.2.1, $\rho_0(y) = -n_0$ for $y \geq 0$ and $\rho_0(y) = -n_0 - y^2$ for $y \leq 0$, where the history at $n = 0$ (assuming the player has elected to continue $-n_0$ times) is adequately summarized by $\xi_0 = Y_0$. At $n = -1$, the choice of stopping at $Y_{-1} = y$ leads to a cost of $(-n_0 - 1)$ while the choice of continuing leads to $Y_0 = y + u_1$ and an expected cost of $E[\rho_0(y + u_1)]$. Thus

$$\rho_{-1}(y) = \min\left[-n_0 - 1, -n_0 - \int_{-\infty}^{-y} (y + u)^2 \varphi(u)\, du\right]$$

and the best action for $Y_{-1} = y$ is to stop or continue depending on which of the two terms in the brackets is smaller. Having evaluated $\rho_{-1}(y)$, one can in principle proceed in the same way to obtain ρ_{-2} and an optimal choice for $n = -2$ as a function of Y_{-2}, etc. For example,

$$\rho_{-2}(y) = \min\left[-n_0 - 2, \int_{-\infty}^{\infty} \rho_{-1}(y + u)\varphi(u)\, du\right].$$

The finite backward induction is generally trivial in principle but difficult to treat analytically. The nontruncated sequential problem can be treated as a limit of the truncated problem, but discrete time stopping problems with infinitely many possible stopping times lead to substantial difficulties which have been discussed effectively in Chow, Robbins and Siegmund [C9]. We shall see that the continuous time stopping problems can be expressed in terms of free boundary problems involving the heat equation and as such provide the possibility of the effective use of the analytic tools of partial differential equations. The justification for the use of continuous time problems lies in the potential power of the analytic methods and the fact that the solutions of these problems approximate the solutions of our discrete time problems.

Although we do not apply backward induction directly, the ideas of backward induction lie behind a good deal of the development in the continuous time stopping problem.

Given a stopping problem as described in § 17.3 let

(17.4.2) $$\rho(y_0, s_0) = \inf_s b(y_0, s_0),$$

where $b(y_0, s_0)$ is the risk associated with a particular stopping time (procedure) S, and the infimum is taken over all procedures S. Note that $\rho(y_0, s_0) \leq d(y_0, s_0)$. But ρ can be defined for all (y, s) with $s \in G$. Since Y is a process of independent increments, it follows that $\rho(y, s)$ also represents the best that can be expected once $Y(s) = y$ is reached, irrespective of how it was reached. Then a characterization of an optimal procedure (under regularity conditions sufficient to imply the existence of an optimal procedure) is described by

(I) S_0: Stop as soon as $\rho(Y(s), s) = d(Y(s), s)$.

Since the optimal procedure S_0 is characterized by the *continuation* set

$$\mathscr{C}_0 = \{(y, s): \rho(y, s) < d(y, s)\}$$

and the *stopping* set

$$\mathscr{S}_0 = \{(y, s): \rho(y, s) = d(y, s)\},$$

we shall restrict our attention to procedures which can be represented by a continuation set $\mathscr{C}$ or its complement the stopping set $\mathscr{S}$.

It is interesting to note that the characterization (I) does not depend on the initial point (y_0, s_0) and thus it yields the solution for all initial points simultaneously, minimizing $b(y, s)$ uniformly for all (y, s).

Under suitable regularity conditions on $d(y, s)$, the solution of the continuous time stopping problem may be approximated by discrete time versions corresponding to a finite sequence of permitted stopping times $(s_1, s_2, \cdots, s_n)$. Since a discrete version permits less choice, the corresponding optimal risk ρ^* is larger and the corresponding optimal continuation set $\mathscr{C}_0^*$ intersects $s = s_i$ on a smaller set than does $\mathscr{C}_0$. As more elements are adjoined to the set of permitted stopping times, ρ^* decreases and the set where $\mathscr{C}_0^*$ intersects $s = s_i$ increases. In this way ρ and $\mathscr{C}_0$ may be derived as limits of monotone sequences.

In the sequential analysis problem the loss due to the discrete time restriction is bounded by the cost of sampling over the discrete time interval. This is so since the statistician may imitate his continuous time solution by stopping as soon as possible after the optimal continuous time solution calls for stopping and by making the same decision. Then the difference in costs between the optimal continuous time procedure and its suboptimal discrete time imitation is simply that of the extra sampling required by the latter.

17.5. Stopping problems: relevance of the heat equation and the free boundary problem. In this section I propose to present a heuristic and brief outline of a discussion explaining (i) the relevance of the heat equation, (ii) why one would anticipate the solution of our optimization problem to be a solution of a *free boundary problem (necessity)*, and (iii) why a solution of the free boundary problem (f.b.p.) is a solution of the optimization problem (*sufficiency*).

A more detailed discussion relating these notions to subparabolic functions is in Chernoff [C10]. Relevant rigorous and more general discussions appear in Bather [B10], Chow, Robbins and Siegmund [C9], Shiraev [S4] and Thompson [T1], and the related concept of excessive function appears in Dynkin [D3]. The necessity part is difficult to phrase precisely and usefully. However, the sufficiency part is easier to treat and more important, for while the necessity condition suggests that we look for solutions of the f.b.p., the sufficiency part tells us when such a solution is the solution of our optimization problem. The difficulty with the necessity part comes from the possibility of singular points which make it difficult to state useful and widely applicable regularity conditions in terms of the given $d(y, s)$ rather than in terms of the unknown solution. When dealing with sufficiency, regularity conditions may be applied to the candidate solution of the f.b.p.

The Wiener process is intimately related to the heat equation. Suppose, for example, that $b(y, s)$ is the expected cost corresponding to an open continuation set $\mathscr{C}$ and stopping cost $d(y, s)$. Then we shall "demonstrate" that

$$(17.5.1) \qquad\qquad \tfrac{1}{2} b_{yy}(y, s) = b_s(y, s), \qquad\qquad (y, s) \in \mathscr{C},$$

while

$$(17.5.2) \qquad\qquad b(y, s) = d(y, s), \qquad\qquad (y, s) \in \mathscr{S}.$$

Suppose $(y, s) \in \mathscr{C}$. Then the probability of stopping between $s + \delta$ and s is $o(\delta)$ and Y changes from $Y(s + \delta)$ to $Y(s)$. Consequently,

$$b(y, s + \delta) = E\{b(Y(s), s) | Y(s + \delta) = y\} + o(\delta)$$
$$= E\{b(y + W\sqrt{\delta}, s)\} + o(\delta),$$

where we use W as a generic $\mathscr{N}(0, 1)$ random variable,

$$b(y, s + \delta) = E\{b(y, s) + W\sqrt{\delta} b_y(y, s) + \tfrac{1}{2} W^2 \delta b_{yy}(y, s) + \cdots\} + o(\delta)$$

$$= b(y, s) + \frac{\delta}{2} b_{yy}(y, s) + o(\delta)$$

and

$$b_s = \tfrac{1}{2}b_{yy}.$$

That $b = d$ on $\mathscr{S}$ is obvious.

Equations (17.5.1) and (17.5.2) apply for an arbitrary procedure. Given $\mathscr{C}$, the solution of (17.5.1) and (17.5.2) is the solution of the *Dirichlet* problem for the heat equation. We require an optimality condition which will help to determine the optimal $\mathscr{C}$ as well as the optimal risk ρ. This extra condition is $\rho_y = d_y$ on the boundary, i.e., the optimal continuation region $\mathscr{C}_0$ and the optimal risk ρ satisfy

(17.5.1*) $$\tfrac{1}{2}\rho_{yy}(y, s) = \rho_s(y, s), \qquad (y, s) \in \mathscr{C}_0,$$

(17.5.2*) $$\rho(y, s) = d(y, s), \qquad (y, s) \in \mathscr{S}_0,$$

(17.5.3) $$\rho_y(y, s) = d_y(y, s) \qquad \text{on the boundary of } \mathscr{C}_0.$$

To demonstrate that ρ is the solution of this *free boundary* or *Stefan* problem let us assume that (y_0, s_0) is a point on a portion of the boundary above which are stopping points and below which are continuation points and that d_y exists at (y_0, s_0). Then since $\rho(y, s_0) = d(y, s_0)$ for $y \geq y_0$, the right-hand derivative $\rho_y^+(y_0, s_0) = d_y(y_0, s_0)$. For $y < y_0$, $\rho(y, s_0) \leq d(y, s_0)$ and hence, $\rho_y^-(y_0, s_0) \geq d_y(y_0, s_0)$. Now we note that

$$\rho(y_0, s_0 + \delta) \leq E\{\rho(y_0 + W\sqrt{\delta}, s_0)\}$$

since the right-hand side corresponds to the risk of the suboptimal procedure where one insists on not stopping between $s_0 + \delta$ and s_0 and proceeding optimally thereafter. But

$$\rho(y_0 + W\sqrt{\delta}, s_0) = \begin{cases} \rho(y_0, s_0) + W\sqrt{\delta}\rho_y^+(y_0, s_0) + o(\sqrt{\delta}), & W > 0, \\ \rho(y_0, s_0) + W\sqrt{\delta}\rho_y^-(y_0, s_0) + o(\sqrt{\delta}), & W < 0, \end{cases}$$

$$E\rho(y_0 + W\sqrt{\delta}, s_0) = \rho(y_0, s_0)$$
$$+ \sqrt{\delta}\left[d_y \int_0^\infty w\varphi(w)\,dw + \rho_y^- \int_{-\infty}^0 w\varphi(w)\,dw \right] + o(\sqrt{\delta})$$
$$= \rho(y_0, s_0) + \sqrt{\delta/2\pi}\{d_y - \rho_y^-\} + o(\sqrt{\delta}).$$

Thus

$$\rho(y_0, s_0 + \delta) - \rho(y_0, s_0) \leq \sqrt{\delta/2\pi}\{d_y - \rho_y^-\} + o(\sqrt{\delta}).$$

Assuming that the difference quotient $[\rho(y_0, s_0 + \delta) - \rho(y_0, s_0)]\delta^{-1}$ is bounded away from $-\infty$, it follows that $d_y - \rho_y^- \geq 0$ which, combined with the results $\rho_y^+ = d_y$ and $\rho_y^- \geq d_y$ establishes (17.5.3).

The free boundary condition (17.5.3) can be replaced by

(17.5.3*) $$\rho_s(y, s) = d_s(y, s) \qquad \text{on the boundary of } \mathscr{C}_0$$

if the optimal boundary is represented by a differentiable function $\tilde{y}(s)$. For then $\rho = d$ along the boundary implies

$$\rho_y \frac{d\tilde{y}}{ds} + \rho_s = d_y \frac{d\tilde{y}}{ds} + d_s$$

and substituting in (17.5.3) yields (17.5.3*). Similarly applying $\rho_y = d_y$ along the boundary yields

$$\rho_{yy} \frac{d\tilde{y}}{ds} + \rho_{ys} = d_{yy} \frac{d\tilde{y}}{ds} + d_{ys}$$

which together with

$$\rho_{ys} = \tfrac{1}{2}\rho_{yyy}$$

gives

(17.5.4) $$\frac{d\tilde{y}}{ds} = \frac{d_{ys} - \rho_{ys}}{\rho_{yy} - d_{yy}} = \frac{d_{ys} - \tfrac{1}{2}\rho_{yyy}}{\rho_{yy} - d_{yy}},$$

where the derivatives are evaluated at $(\tilde{y}(s), s)$.

Returning to the f.b.p. the following question arises. *Is a solution of the f.b.p. necessarily a solution of the optimization problem*? The answer is, "Yes", providing certain additional conditions are satisfied. That additional conditions are required is clear from the following considerations. Suppose that $(u, \mathscr{C})$ is a solution of both the f.b.p. ($\tfrac{1}{2}u_{yy} = u_s$ on $\mathscr{C}$, $u = d$ on $\mathscr{S}$ and $u_y = d_y$ on the boundary) and the optimization problem ($\mathscr{C}_0 = \mathscr{C}$, $\rho = u$). If the problem is modified by sharply decreasing d below u on part of $\mathscr{C}_0$, then $(u, \mathscr{C})$ remains a solution of the f.b.p. but the solution of the optimization problem changes. If d is sharply decreased on a small part of the stopping set near the boundary of $\mathscr{C}$ the optimal continuation region should be enlarged, but here again $\mathscr{C}$ still corresponds to a solution of the f.b.p.

These examples suggest sufficiency conditions which may be paraphrased to state that *if one cannot trivially improve on u* (as was possible in the above counterexamples), *then* $\rho = u$ *and* $\mathscr{C}$ *is optimal*. Those conditions would be that

(i) $u(y, s) \leq d(y, s)$, and
(ii) $u(y, s; s') \geq d(y, s)$ for $(y, s) \in \mathscr{S}$ and $s \geq s'$,
where the notation

$$h(y, s; s') = E\{h(y + W\sqrt{s - s'}, s')\} \quad \text{for} \quad s \geq s'$$

represents the expectation of $h(Y(s'), s')$ given $Y(s) = y$. In other words, $u(y, s; s')$ represents the risk to the statistician when he observes $Y(s) = y$ and constrains himself to continue till s' and pay $u(Y(s'), s')$.

To establish the sufficiency of these conditions, one assumes that a better procedure exists which can be approximated by a stopping procedure which permits stopping only on a discrete set of times. But a backward induction invoking these conditions shows that at each stage the risk of this new procedure is at least u, leading to a contradiction.

Applying condition (ii) above, one may derive the following theorem.

THEOREM 17.5.1 (Sufficiency). *If* $(u, \mathscr{C})$ *is a solution of the f.b.p. where* $\mathscr{C}$ *is a continuation set and u and d have bounded derivatives up to third order and*

(i) $u(y, s) \leqq d(y, s)$ *and*

(ii) $\frac{1}{2} d_{yy} \geqq d_s$ *on* $\mathscr{S}$,

then $(u, \mathscr{C})$ *is a solution of the optimization problem under the proviso that the optimal risk can be approximated by the risk of a procedure where stopping is restricted to a finite number of discrete times.*

In some applications Theorem 17.5.1 is not powerful enough because some of the conditions break down as s approaches its lower limit $s^{(0)}$ (possibly $-\infty$). In that case it suffices to establish the conditions of the theorem for problems which terminate by $s = s^* > s^{(0)}$ and then to invoke some supplementary condition which implies

$$(17.5.5) \qquad \limsup_{s \to s^{(0)}} |u(y, s) - \rho(y, s)| = 0.$$

This condition is unnecessarily strong as are many of the boundedness conditions. For example, if b and d and their derivatives are $o[\exp(Ky^2)]$ for each $K > 0$, proofs will go through with minor modification.

17.6. Bounds and asymptotic expansions for solutions. The continuous time stopping problem, Example 17.3.2, has a trivial solution. One way to see it is to observe that it is invariant under the transformation $Y^* = aY, s^* = a^2 s$. We can also see this by proposing $\mathscr{C}_0 = \{(y, s): y < 0, s > 0\}$ and

$$\rho(y, s) = -s, \qquad\qquad y \geqq 0, s \geqq 0,$$
$$\rho(y, s) = -y^2 - s, \qquad\qquad y < 0, s \geqq 0.$$

The pair $(\rho, \mathscr{C}_0)$ is a solution of the f.b.p. since $\rho = d$ and $\rho_y = d_y$ for $y = 0$. The sufficiency conditions of Theorem 17.5.1, slightly modified to allow for the fact that ρ and d are not bounded, apply.

Generally, stopping problems are not so easily solved. It is useful to derive bounds on ρ and $\mathscr{C}_0$. Bather [B6] exploited the following techniques very effectively.

First, let $u(y, s)$ be an *arbitrary* solution of the heat equation. Let $\mathscr{B}$ be the set on which $u(y, s) = d(y, s)$. If $\mathscr{B}$ is the boundary of a continuation set $\mathscr{C}$, the risk for the procedure defined by the continuation set $\mathscr{C}$ is $b(y, s) = u(y, s)$ on $\mathscr{C}$ and $b(y, s) = d(y, s)$ on $\mathscr{S}$. But then $\rho(y, s) \leqq b(y, s)$. Thus if (y_0, s_0) is a point of $\mathscr{C}$ where $u < d$, then $\rho(y_0, s_0) < d(y_0, s_0)$ and (y_0, s_0) is a continuation point for the optimal procedure.

We illustrate with a new stopping problem. This problem is important in a control theory application and it is discussed in greater detail in [C10]. For our purposes it provides an illustration of the main idea with relatively little analysis.

PROBLEM 17.6.1. In this problem

$$d(y, s) = \begin{cases} y & \text{for} \quad s = 0, \quad y \geq 0, \\ s^{-1} & \text{for} \quad s > 0, \quad y > 0, \\ 0 & \text{for} \quad s \geq 0, \quad y \leq 0, \end{cases}$$

and stopping is enforced when $Y(s) \leq 0$ or $s = 0$.

Note that if s is large, the chances of obtaining $Y(s) = 0$ (and zero cost) before $s = 0$ is large and so one is encouraged to continue unless Y is large. If s is small, the cost of stopping, s^{-1}, is large compared to the cost of waiting until $s = 0$ (which is approximately Y) and one is encouraged to continue unless Y is large. Thus one expects $\mathscr{C}_0$ to have a boundary which is high for s large and s small.

Take $u_1(y, s) = y$ which is a solution of the heat equation. Since $u_1(y, s) = d(y, s)$ for $y = s^{-1}$, let $\mathscr{C}_1 = \{(y, s) : 0 < y < s^{-1}, s > 0\}$. $u_1 = d$ on the boundary of $\mathscr{C}_1$, the upper part of which is $y_1(s) = s^{-1}$. Because $u_1(y, s) < s^{-1}$ in $\mathscr{C}_1$, every point of $\mathscr{C}_1$ is in $\mathscr{C}_0$ and $y_1(s) = s^{-1}$ is a lower bound for the optimal boundary $\tilde{y}(s)$, i.e.,

(17.6.1) $\tilde{y}(s) \geq y_1(s) = s^{-1}$.

Moreover, $\rho(y, s) \leq y$ in $\mathscr{C}_1$.

As another illustration of the method, take

$$u_2(y, s) = Ays^{-3/2}\varphi(ys^{-1/2}),$$

which is a solution of the heat equation that can be written as $A\alpha\varphi(\alpha)s^{-1}$, where $\alpha = ys^{-1/2}$. Then $u_2 = d$ where

$$A\alpha\varphi(\alpha) = 1$$

as well as for $y = 0$. For $A > [\max \alpha\varphi(\alpha)]^{-1}$, there are two positive values α_1 and α_2, so that $A\alpha\varphi(\alpha) < 1$ for $0 < \alpha < \alpha_1$ and $\alpha_2 < \alpha < \infty$ while $A\alpha\varphi(\alpha) > 1$ for $\alpha_1 < \alpha < \alpha_2$. Then we may take $y_2(s)$ to be either $\alpha_1 s^{1/2}$ or $\alpha_2 s^{1/2}$, and $\mathscr{C}_2 = \{(y, s) : 0 < y < y_2(s)\}$. Then u_2 corresponds to the risk of the procedure defined by $\mathscr{C}_2$. Moreover $u_2 \leq d$ for $y \leq \alpha_1 s^{1/2}$. Thus $\alpha_1 s^{1/2}$ is a lower bound for the optimal boundary. We obtain the best bound of this form by taking $A = [\max \alpha\varphi(\alpha)]^{-1}$ which is attained at $\alpha = 1$. Thus

(17.6.2) $\tilde{y}(s) \geq s^{1/2}$.

An interesting aspect of using $u_2 = [\varphi(1)]^{-1}\alpha\varphi(\alpha)s^{-1}$ is that it provides a solution of the f.b.p. However, it is not our optimal solution since the sufficiency conditions fail to hold in the neighborhood of $(y, s) = (0, 0)$.

The method just applied yields an inner bound on $\mathscr{C}_0$. We now describe a method of finding an outer bound. This method may be paraphrased as follows. Let $u(y, s)$ be a solution of the heat equation. It may not be optimal for our problem, but find the problem for which it is optimal. By comparing these two problems we derive bounds.

More specifically, let $\mathscr{B}$ be the set on which $u_y(y, s) = d_y(y, s)$. Suppose $\mathscr{B}$ is the boundary of a continuation set $\mathscr{C}$. If $u \neq d$ on $\mathscr{B}$ let $h(s) = u(y, s) - d(y, s)$ along the boundary $\mathscr{B}$ and let $d^*(y, s) = d(y, s) + h(s)$. Then $(u, \mathscr{C})$ is a solution of the f.b.p. for $d^*(y, s)$. Suppose that $(u, \mathscr{C})$ is also a solution of the optimality problem for d^* and $h(s) \leq 0$ for $s \leq s_2$ and $h(s_2) = 0$. Then the modified problem is a more advantageous problem than the original for $s = s_2$ and

$$\rho(y, s_2) \geq u(y, s_2).$$

If (y_2, s_2) is a stopping point for the modified problem,

$$\rho(y_2, s_2) \geq u(y_2, s_2) = d^*(y_2, s_2) = d(y_2, s_2)$$

and (y_2, s_2) is a stopping point for the original problem.

To illustrate,

$$u_3(y, s) = y - B e^{a^2 s/2} \sinh ay$$

is a solution of the heat equation for which $u_{3y} = d_y = 0$ when $y = y_3(s)$, which is determined, for $a > 0, B > 0, 0 < aB < 1$, by

$$1 - Bae^{a^2 s/2} \cosh ay = 0.$$

For $s = 0$, $u_3 - d \leq 0$. Along the boundary, $y = y_3(s)$, $u_3 - d$ takes on negative values for small positive s. The smallest positive value s_3 of s (if any) where $u_3 - d$ vanishes at $(y, s) = (y_3(s_3), s_3)$ is described by

$$y - Be^{a^2 s/2} \sinh ay = s^{-1}.$$

Any pair of parameters (a, B) which yield such a pair (y_3, s_3) may be used, and the corresponding point (y_3, s_3) is a point of $\mathscr{S}_0$.

A minimization argument may be used to find the best pair (a, B) for each s_3 as the solution of $\partial(u_3 - d)/\partial s = 0$ and this leads to the outer bound

$$\tilde{y}(s) \leq \tilde{y}_3(s),$$

where

(17.6.3)
$$\frac{2^{1/2}}{s}(\tilde{y}_3 - s^{-1})^{1/2} = \tanh\left\{\frac{2^{1/2}\tilde{y}_3}{s}(\tilde{y}_3 - s^{-1})^{-1/2}\right\},$$

which is very close to the inner bound for s close to zero.

The effective use of these techniques requires the ability to generate convenient solutions of the heat equation. This is also true of the related approach to approximating the optimal boundary which consists of finding asymptotic expansions for the risk and boundary near distinguished points of s; these distinguished points are typically the endpoints of the range of interest. For example, $s = 0$ and ∞ are such points in the sequential analysis problem and Problem 17.6.1.

One important class of solutions of the heat equation used in generating expansions is that generated by "sources of heat" along a vertical ($s = $ constant) line. Thus

(17.6.4)
$$u_0(y, s) = s^{-1/2}\varphi(\alpha), \quad \alpha = y/\sqrt{s},$$

represents a point source of heat at $(y, s) = (0, 0)$ and yields a solution of the heat equation for $s > 0$. Similarly, functions of the form

$$(17.6.5) \qquad u(y, s) = \int \frac{1}{\sqrt{s}} \varphi\left(\frac{y - y'}{\sqrt{s}}\right) h(y') \, dy' = \int h(y + w\sqrt{s})\varphi(w) \, dw$$

satisfy the heat equation.

In particular, if $h(y) = y^r/r!$ for $y > 0$ and 0 otherwise, then

$$u(y, s) = s^{r/2} G_r(\alpha),$$

where

$$G_r(\alpha) = P_r(\alpha)\Phi(\alpha) + Q_r(\alpha)\varphi(\alpha)$$

and the P_r and Q_r are easily derived polynomials for integer values of r. These are special solutions of the heat equation of the form

$$u(y, s) = s^{r/2} H_r(\alpha),$$

where H_r satisfies the ordinary differential equation

$$H_r''(\alpha) + \alpha H_r'(\alpha) = r H_r(\alpha)$$

which gives *hypergeometric functions*. Note that H_r' is a candidate for H_{r-1}. For negative integer values the H_r involve $\varphi(\alpha)$ and $J(\alpha) = \varphi(\alpha) \int_0^\alpha [\varphi(t)]^{-1} \, dt$ with coefficients which are polynomials in α.

Returning to the normalized sequential problem, Example 17.3.1, we approximate near $s = 0$ by an even solution of the form

$$u(y, s) = c_0 s^{-1} H_{-2}(\alpha) + c_1 s^{1/2} H_1(\alpha) + c_2 s^2 H_4(\alpha) + \cdots,$$

where the leading term is suggested by the s^{-1} component of d. For $s \approx \infty$ one may use an even solution of the form

$$u(y, s) = K_1 s^{-1/2} \varphi(\alpha) + \int_0^\infty [\varphi(\alpha - \alpha') + \varphi(\alpha + \alpha') - 2\varphi(\alpha)] h(|s^{1/2}\alpha'|) \, d\alpha',$$

where

$$h(x) \approx \frac{K_2(\log x^2)}{x^2} \{1 + R_1(\log x^2)^{-1} + R_2(\log x^2)^{-2} + \cdots\}$$

as $x \to \infty$ and the R_i are polynomials in $\log[\log x^2]$. This form is related to that of (17.6.5). The first term is a point source of heat at $(y, s) = (0, 0)$. Then the remainder is the limit as $A \to \infty$ of

$$\int_{-A}^A [\varphi(\alpha - \alpha') - \varphi(\alpha)] h(|s^{1/2}\alpha'|) \, d\alpha'$$

and represents a symmetric distribution of heat along the $s = 0$ axis which is compensated for by an equivalent (but negative) point source at $(y, s) = (0, 0)$. This form, alternative to (17.6.5), is more convenient, when $h(x)$ behaves like

$(\log x^2)/x^2$ as $x \to \infty$, for deriving asymptotic expansions for u as $s \to \infty$ and α is fixed or grows slowly with s. The choice of h presented above was suggested by matching expansions for the two boundary conditions termwise.

That these expansions yield asymptotic approximations for the boundary and risk functions is proved by taking a finite number of terms in the expansions and varying the coefficient of the last (highest order) term. The bounding methods described above are used to show that these adjusted expansions yield upper and lower bounds.

The following expansions for the optimal boundary $\tilde{y}(s)$ were derived in [B8], [C12] for the normalized sequential analysis problem Example 17.3.1:

$$(17.6.6) \quad \tilde{a}(s) = \tilde{y}(s)s^{-1/2} \sim \{\log s^3 - \log 8\pi - 6(\log s^3)^{-1} + \cdots\}^{1/2} \quad \text{as} \quad s \to \infty,$$

$$(17.6.7) \quad \tilde{a}(s) = \tilde{y}(s)s^{-1/2} \sim \frac{1}{4}s^{3/2}\left\{1 - \frac{1}{12}s^3 + \frac{7}{15 \cdot 16}s^6 - \cdots\right\} \quad \text{as} \quad s \to 0.$$

These expansions could be expressed in a somewhat more "familiar" form by regarding the normalized problem as one where $k = c = \sigma = 1$. Then

$$Y = \frac{\mu_0\sigma_0^{-2} + X(t)}{\sigma_0^{-2} + t} \quad \text{and} \quad s^{-1} = \sigma_0^{-2} + t.$$

The prior distribution has the same effect as would be derived by starting with the ideal case of no prior knowledge and, prior to our observation, having observed the Wiener process reach $X_0 = \mu_0\sigma_0^{-2}$ in time $t_0 = \sigma_0^{-2}$. Then $Y = X'/t'$ and $s^{-1} = t'$, where

$$X' = X_0 + X(t) = Y/s \quad \text{and} \quad t' = t_0 + t = s^{-1}$$

expresses the position of the Wiener process originating from (X_0, t_0) instead of from the origin. In the (X', t')-space all normalized versions of the problem are the same except for the starting point $(X_0, t_0) = (\mu_0\sigma_0^{-2}, \sigma_0^{-2})$ determined by the prior. As the prior information decreases, so does $t_0 = \sigma_0^{-2}$ and (X_0, t_0) moves closer to the origin (unless μ_0 changes rather extremely). At any time t' represents the total information or precision, including the prior, accumulated to date. That the current estimate of μ should be $Y = X'/t'$ is not surprising.

In the (X', t') scale, the above expansions yield

$$(17.6.6') \qquad \tilde{x}' \approx \{t'[-3\log t' - \log 8\pi]\}^{1/2} \quad \text{as} \quad t' \to 0, \qquad\qquad s \to \infty,$$

$$(17.6.7') \qquad \tilde{x}' \approx 1/4t' \qquad\qquad\qquad \text{as} \quad t' \to \infty, \qquad\qquad s \to 0.$$

At the same time these expansions yield estimates on the Bayes risk which are of order

$$(17.6.6'') \qquad \mathscr{B}(y, s) \approx K_1 s^{-1/2}\varphi(\alpha) + O[s^{-1}(\log s)^2] \quad \text{as} \quad t' \to 0, \qquad s \to \infty,$$

$$\approx K_1 t'^{1/2}\varphi(\alpha) + O[t'(\log t')^2]$$

$$(17.6.7'') \qquad \mathscr{B}(y, s) \approx s^{-1} \approx t' \qquad\qquad\qquad \text{as} \quad t' \to \infty, \qquad s \to 0,$$

where $\alpha = y/\sqrt{s} = x'/\sqrt{t'}$.

As we shall soon see, the case where $c \to 0$ in the unnormalized case corresponds to $t' \to 0$. In that case the constant K_1 can be decomposed into two parts, one representing the contribution of sampling and the other that of the expected cost of error. These two costs are comparable.

The non-Bayesian would like to compute the risks $R(\mu)$ for the procedures used and we have merely presented the average

$$\mathscr{B}(y, s) = \int R(\mu)n(\mu; y, s)\, d\mu.$$

But this equation may be regarded as a form of Laplace transform which should be inverted. This has not yet been done.

It is a simple matter to transform these expressions to comparable ones in the nonnormalized form. In (17.6.6), (17.6.7) and their variations, $s, \tilde{y}, \mathscr{B}, \tilde{\alpha}, t', \tilde{x}$ must be replaced by $s^* = a^2 s$, $\tilde{y}^* = a\tilde{y}$, $\mathscr{B}^* = (\mathscr{B} + c\sigma_0^{-2}\sigma^2)b$, $\tilde{\alpha}^* = \tilde{\alpha}$, $t^* = a^{-2}t'$, $\tilde{x}^* = a^{-1}\tilde{x}'$, where $a = c^{-1/3}k^{1/3}\sigma^{-2/3}$, $b = c^{-1/3}k^{-2/3}\sigma^{-2/3}$, $t' = \sigma_0^{-2} + t\sigma^{-2}$ and $\tilde{x}' = \mu_0\sigma_0^{-2} + \tilde{x}\sigma^{-2}$. The leading term in the Bayes risk is $O(s^{-1/2})$ in the normalized problem. In the unnormalized problem this leads to $\mathscr{B}(\mu_0, \sigma_0^{-2}) \sim c^{2/3}$ which is a substantial improvement over that of the fixed sample size case where the risk is of order $c^{1/2}$.

17.7. Discrete time problems revisited.
Having obtained approximations and expansions for the continuous time problem, let us raise a few questions. How do we compute a good approximation to the continuous version where the asymptotic expansions do not fit well and where our ingenuity did not provide good bounds?

I propose that we perform a backward induction to solve the discrete time problem and use that solution as an approximation to the continuous time problem. Indeed the discrete time problem is naturally truncated. It is easy to see that for t sufficiently large, the expected cost of error, $s^{1/2}\psi(ys^{-1/2})$ is less than c for all y. Hence it cannot be decreased enough to compensate for the cost of an extra observation. Since the problem is truncated, the backward induction involves a finite number of steps which can be carried out by numerical integration.

But we have become rather circular. We introduced the continuous version to avoid backward induction, and now we solve that by backward induction. This is not quite the situation. First our continuous time version has given us general insights on the limiting case as $c \to 0$. Second, if we had a good approximation to the difference between the solution of the continuous time problem and that of the discrete version, we could solve one discrete version numerically and use its solution, properly adjusted, to approximate the continuous one *as well as the other discrete versions. Thus one numerical backward induction does the work for all.*

We shall outline the derivation of a rather remarkable result [C13]. If the problem is transformed to the (X', t')-space in normalized form, i.e., $t' = c^{2/3}k^{-2/3}\sigma^{4/3}$ $\cdot(\sigma_0^{-2} + t\sigma^{-2})$, $X' = c^{1/3}k^{-1/3}\sigma^{2/3}(\mu_0\sigma_0^{-2} + X\sigma^{-2})$, the stopping times are separated by intervals of length $\delta_{t'} = c^{2/3}k^{-2/3}\sigma^{-2/3}$ and *the difference between the discrete solution $\tilde{x}'_\delta$ and the continuous solution $\tilde{x}'$ at the upper boundary is expressed* by

(17.7.1) $$\tilde{x}' - \tilde{x}'_\delta \approx 0.58\sqrt{\delta_{t'}} = 0.58c^{1/3}k^{-1/3}\sigma^{-1/3}$$

as long as t' is bounded away from zero and ∞. This is proved by showing that when the intervals δ_s between successive values of s are small and approximately constant locally, then $\tilde{y} - \tilde{y}_\delta \approx 0.58\sqrt{\delta_s}$, and making the necessary transformations to the (x', t') scale.

To prove this result one imagines a statistician located near a point at the optimal boundary (y_0, s_0). Viewing the region with a microscope, the problem can be separated into two parts, one of which is irrelevant and the other resembles our trivial Example 17.3.2. But for the statistician who is constrained to stop at discrete times, it resembles the discrete version of Example 17.3.2, namely Example 17.2.1. Then the difference between the solutions of our continuous and discrete problems are related, with a scale change, to those between the discrete and continuous versions of Example 17.3.2. The continuous version was trivial with optimal boundary given by $\tilde{y} = 0$. The discrete version is subject to analysis and yields an optimal boundary $\tilde{y}_n \to -0.58$ as $n \to \infty$. Applying the relations the results follow.

We outline a few of the details. First, for $s > s_0$,

(17.7.2) $$d(y, s; s_0) = E[d[Y(s_0), s_0]| Y(s) = y]$$

represents the expected loss given $Y(s) = y$, if stopping with resulting loss $d(Y, s)$ is permitted only at s_0. Hence $d(y, s; s_0)$ satisfies the heat equation for $s > s_0$. In a problem with stopping risk $d^*(y, s) = d(y, s; s_0)$ it is irrelevant what stopping plan is used. Thus we need consider only $d^{**}(y, s) = d(y, s) - d(y, s; s_0)$ which has the same solution as $d(y, s)$ with optimal risk $\rho^{**}(y, s) = \rho(y, s) - d(y, s; s_0)$.

Suppose that $\mathscr{C}_0$ is below $\mathscr{S}_0$ at the boundary point (y_0, s_0). If we expand about (y_0, s_0) using the fact that $d_y(y, s_0; s_0) = d_y(y, s_0)$ and $d_s(y, s_0; s_0) = \frac{1}{2}d_{yy}(y, s_0; s_0) = \frac{1}{2}d_{yy}(y, s_0)$, we have, for $s \geqq s_0$,

$$d(y, s) - d(y, s; s_0) = d(y_0, s_0) + (s - s_0) d_s + (y - y_0) d_y + \frac{1}{2}(y - y_0)^2 d_{yy} + \cdots$$
$$- [d(y_0, s_0) + \frac{1}{2}(s - s_0) d_{yy}$$
$$+ (y - y_0) d_y + \frac{1}{2}(y - y_0)^2 d_{yy} + \cdots],$$

(17.7.3) $$d(y, s) - d(y, s; s_0) \approx \delta_s [d_s - \frac{1}{2} d_{yy}]\left(\frac{s - s_0}{\delta_s}\right).$$

For $y < y_0, s = s_0$,

$$\rho(y, s_0) - d(y, s_0; s_0) = \rho(y, s_0) - d(y, s_0) \approx \frac{1}{2}(y - y_0)^2 [\rho_{yy} - d_{yy}].$$

But $\frac{1}{2}\rho_{yy} = \rho_s = d_s$ and hence,

(17.7.4) $$\rho(y, s_0) - d(y, s_0; s_0) \approx \delta_s [d_s - \frac{1}{2} d_{yy}]\left(\frac{y - y_0}{\sqrt{\delta_s}}\right)^2,$$

where $d_s - \frac{1}{2} d_{yy} < 0$ on the boundary. Thus the stopping problem with $d^{**}(y, s) = d(y, s) - d(y, s; s_0)$ for $s > s_0$ and $d^{**}(y, s) = \rho(y, s_0) - d(y, s_0; s_0)$ for $s = s_0$

is locally equivalent, except for a scale factor, to Example 17.3.2 or Example 17.2.1, depending on whether we permit stopping continuously or at intervals of length δ_s. These problems relate to our original problem, except that once s_0 is reached, we are assumed to proceed optimally (and continuously).

Now an argument is required to show that there is relatively little difference in boundary at $s = s_0 + n\delta_s$, where n is large but $n\delta_s$ is small, for the solution of two problems. One is the original discrete problem with stopping cost $d(y, s)$. The other is the modified problem where only discrete stopping is allowed for $s > s_0$ but continuous stopping is permitted for $s \leqq s_0$.

The details appear in [C13].

Ordinarily the numerical work involved in backward induction is time-consuming and the programming is moderately complicated. I have applied a slight modification which reduces the programming labor considerably.

This consists of replacing the Gaussian $Y(s)$ process by a discrete time-discrete step process. Let $Y(s)$ be a process which is defined for integer values of s, subject to

$$Y(s - 1) = Y(s) \pm 1 \quad \text{with probability } 1/2$$

independent of the past, i.e., $[Y(s), Y(s + 1), \cdots]$. Then $Y(s)$ has independent increments with mean 0 and variance 1 per unit change in s. Given a stopping problem defined by $d(y, s)$, we can imitate it by a suitable transformation in the (y, s) scale, i.e., $y^* = ay$, $s^* = a^2s$, and applying the discrete process our backward induction reduces to

$$\rho(y, s) = \min[d(y, s), [\rho(y + 1, s - 1) + \rho(y - 1, s - 1)]/2],$$

which is relatively simple algorithm. Using this algorithm leads to a variation of the correction for discreteness (17.7.1), where 0.58 is replaced by 0.5 [C20].

18. Bandit problems. A stopping problem of special interest is the following.

PROBLEM 18.1. (One-armed bandit). Let $X(t)$ be a Wiener process with unknown drift μ per unit time, i.e., $E\{dX(t)\} = \mu \, dt$, $\text{var}\{dX(t)\} = \sigma^2 \, dt$, σ^2 known.

One is permitted to stop at any time t, $0 \leqq t \leqq n_0$, and receive a payoff $X(t)$. The unknown parameter μ has the prior distribution $\mathcal{N}(\mu_0, \sigma_0^2)$. What is the optimal stopping procedure?

This problem is the continuous time version of a discrete time problem where a player engages in a game. After the ith play he receives a reward X_i (possibly negative) and decides whether to continue playing. He is permitted to play at most n_0 games (n_0 a specified number). As he plays, his losses or gains cumulate and simultaneously he receives information about whether the game is a profitable one or not.

Since $E(X_1 + \cdots + X_N) = E(N\mu)$ the problem can be phrased in terms of finding a stopping rule N subject to $N \leqq n_0$ which maximizes $E(N\mu)$.

If the player knew μ he would play n_0 times for $\mu > 0$ and 0 times for $\mu < 0$. If he plays N times, his expected loss due to ignorance is $-N\mu$ if $\mu < 0$ and $(n_0 - N)\mu$ if $\mu > 0$.

The methods of § 17 apply here. We have the same posterior distribution as in § 17.3 where $\mathscr{L}(\mu|X(t'), 0 \leq t' \leq i) = \mathscr{N}(Y(s), s)$, $Y(s) = [\mu_0\sigma_0^{-2} + X(t)\sigma^{-2}]/(\sigma_0^{-2} + t\sigma^{-2})$, and $s^{-1} = (\sigma_0^{-2} + t\sigma^{-2})$. Here s varies from $s_0 = \sigma_0^2$ to $s_1 = (\sigma_0^{-2} + n_0\sigma^{-2})^{-1}$. The loss on stopping at $(Y(s), s)$ is $-X(t) = \sigma^2[\mu_0\sigma_0^{-2} - Y(s)/s]$ which is a linear function of $-Y(s)/s$.

By applying the transformation $s^* = s/s_1$, $Y^* = Y/\sqrt{s_1}$, we normalize so that $s_1^* = 1$. In this way Problem 18.1 is reduced to a normalized stopping problem with

(18.1) $$d(y^*, s^*) = -y^*/s^* \quad \text{for} \quad s^* \geq 1$$

with stopping enforced at $s^* = 1$.

Chernoff and Ray [C14] derived expansions using the methods of § 17.6. It is convenient to express the solution in terms of a functional relationship between

(18.2) $$\beta = \Phi[\tilde{y}^*(s^*)/\sqrt{s^*}] \quad \text{and} \quad t^* = s^{*-1}.$$

Here t^* represents the ratio of the information (on μ) that is presently available to that which would be ultimately available if we continue till forced to stop. At the same time β represents a *nominal significance level corresponding to the test of the hypothesis* $\mu = 0$ since $Y(s)/\sqrt{s}$ is the number of standard deviations that our current estimate of μ is away from zero. Thus our procedure may be described as a series of significance tests of $H : \mu = 0$. We stop when we decide $\mu < 0$. The significance level of these tests changes with t^*. For small t^* the optimal procedure insists on a *strict* significance level. In general,

$$\beta = \Phi(\tilde{y}(s)/\sqrt{s}),$$

where $\beta \approx 2t^*$ for t^* small and $\beta \to 1/2$ for $t^* \to 1$. The nominal significance level β is tabulated in Table 4 as a function of t^*. In addition three other functions are included. These are

(18.3) $$\alpha_1(t^*) = -\tilde{y}(s)/\sqrt{s},$$

which is the number of standard deviations corresponding to β and $-\tilde{y}^*(s^*)$. One more function α_2, to which we shall refer later, is also included.

Without going into the details we may remark on the risk associated with the optimal procedure. A certain gain would be expected in the ideal situation where nature selects μ according to $\mathscr{N}(\mu_0, \sigma_0^2)$ and a spy informs the gambler whether $\mu > 0$ or $\mu < 0$, so that he can select $t = 0$ or n_0 accordingly. One anticipates that as $n_0 \to \infty$, the *expected loss due to ignorance* will become large, for if a million plays are permitted, it is important to invest heavily enough to be reasonably sure that $\mu < 0$ before stopping. On the other hand, if μ is substantially negative, evidence cumulates exponentially and one anticipates a relatively small loss due to ignorance. Indeed, for the optimal procedure this loss is of the order of magnitude of $\sigma(\log n_0)^2$.

The one-armed bandit problem is relevant in several applications. One is that of clinical trials. A new treatment is proposed which is to be compared with one of known efficacy. It is anticipated that a large number n_0 of patients will have to be treated eventually by one of these treatments. The value of a scheme of allocation

HERMAN CHERNOFF

TABLE 4

Tabulations relevant to solutions of bandit problems

t^*	$-\bar{y}^*(t^{*-1})$	$\tilde{\beta}(t^*)$	$\alpha_1(t^*)$	$\alpha_2(t^*)$
0.0001	354.70	1.95(−4)	3.547	3.891
0.0002	237.87	3.84(−4)	3.364	3.719
0.0005	138.77	9.58(−4)	3.103	3.481
0.001	91.80	1.85(−3)	2.903	3.291
0.002	59.54	3.68(−3)	2.684	3.090
0.005	33.52	8.89(−3)	2.370	2.807
0.01	22.45	0.0214	2.245	2.576
0.02	13.14	0.0316	1.852	2.326
0.04	7.72	0.0613	1.544	2.054
0.06	6.09	0.0680	1.491	1.880
0.08	4.91	0.082	1.389	1.751
0.10	4.07	0.099	1.287	1.645
0.15	2.82	0.137	1.094	1.439
0.20	2.18	0.165	0.975	1.282
0.25	1.76	0.189	0.882	1.150
0.30	1.47	0.210	0.805	1.036
0.40	1.07	0.248	0.680	0.842
0.50	0.818	0.281	0.579	0.675
0.60	0.627	0.314	0.486	0.524
0.70	0.477	0.345	0.399	0.385
0.80	0.347	0.378	0.311	0.253
0.85	0.283	0.397	0.260	0.189
0.90	0.222	0.417	0.210	0.126
0.92	0.194	0.426	0.186	0.100
0.94	0.157	0.440	0.152	0.753
0.96	0.130	0.449	0.128	0.502
0.98	0.087	0.466	0.086	0.251
0.99	0.063	0.475	0.062	$\cdots$
1.00	0.000	0.500	0.000	0.000

1. $\bar{y}^*(s^*)$ designates solution of normalized one-armed bandit problem.
2. $\tilde{\beta}$ is nominal significance level for one-armed bandit problem.
3. α_1 is corresponding number of standard deviations.
4. α_2 is function used in the approximation to $g(s_1^*, s_2^*)$ which represents the solution of the two-armed bandit problem.
$$t^* = s^{*-1}, \alpha_1 = -(t^*)^{1/2}\bar{y}^*(t^{*-1}), \tilde{\beta} = \Phi(-\alpha_1).$$
For $0.99 < t^* < 1.00$ use $\alpha_1 = 0.639(1 - t^*)^{1/2}$.

of treatments is the total number of patients successfully treated. If the standard treatment has known probability p_0 and the new one has unknown probability p, and N patients are given the new treatment, the loss due to ignorance will be

$$-N(p - p_0) \quad \text{if} \quad p - p_0 < 0$$

and

$$(n_0 - N)(p - p_0) \quad \text{if} \quad p - p_0 > 0.$$

Since trials on the standard treatment give no useful information on p, and p_0 is known, it is only reasonable that the new treatment be tried until it is determined to be not good enough. This problem is of the same form as the one-armed bandit except that $p - p_0$ replaces μ and the observations are successes or failures and not normally distributed. If $\hat{p}_n$ is the proportion of successes in n trials, $\mathscr{L}(\hat{p}_n - p_0) \approx \mathscr{N}(p - p_0, p_0(1 - p_0)/n)$ when p is close to p_0, and then $\hat{p}_n - p_0$ corresponds to $\bar{X}_n$ and $p_0(1 - p_0)$ to σ^2 in the discrete version of the bandit problem.

A number of serious questions concerning ethics and practicality arise. One issue is that typically n_0 is not well known. Fortunately this is not a serious problem, since n_0 can change by a large factor without having a major effect on the optimal procedure.

Many plans in practice have theoretical and practical drawbacks. Theoretically, they tend to have losses of the order of $n_0^{1/3}$ or $n_0^{1/2}$, while the losses for the optimal plan are of the order of $(\log n_0)^2$.

Suppose each of the trials leads to failure. How many successive failures are needed before switching back to the standard? The optimal solution gives a number of order $\log n_0$ but some typical schemes give numbers of order $n_0^{1/3}$ or $n_0^{1/2}$.

Another application, one which led to the solution in [C14], is the rectified sampling inspection problem. Here a lot of items is available for inspection. There is a cost of inspection per item sampled. Any bad item inspected is detected and replaced by a good item. After some inspection, the rest of the lot is passed. If a bad item is passed, it will lead to a cost. Depending on the proportion of bad items in the lot, one should have either 100% inspection or 0 inspection. Not knowing this proportion, one samples at random, replacing defective items, until one decides that the lot is good enough not to require further inspection or until one has inspected the whole lot.

Finally another application to an experimental design problem will be considered in § 20.

A related problem is the *two-armed bandit* problem. Let X_1 and X_2 be two normally distributed variables with unknown means μ_1 and μ_2 and common known variance σ^2. A specified total of n_0 observations must be allocated to these variables. The player keeps

$$(18.4) \qquad X_{11} + \cdots + X_{1N_1} + X_{21} + \cdots + X_{2N_2}, \qquad N_1 + N_2 = n_0,$$

where $X_{i1}, X_{i2}, \cdots, X_{iN_i}$ are the observations on X_i, $i = 1, 2$. Two independent normal prior distributions are given for μ_1 and μ_2. Having taken m_i observations on X_i, $i = 1, 2, m_1 + m_2 < n_0$, how should one decide between X_1 and X_2 for the next observation? Note that

$$(18.5) \quad E(X_{11} + \cdots + X_{1N_1} + X_{21} + \cdots + X_{2N_2}) = E(N_1\mu_1 + N_2\mu_2)$$

$$= n_0\mu_1 + E[N_2(\mu_2 - \mu_1)]$$

and hence one is concerned with minimizing $E[N_2(\mu_1 - \mu_2)]$.

The two-armed bandit problem where the prior distribution on μ_1 has 0 variance reduces to the one-armed bandit problem. One expects a strong relationship between the behavior of the solution of the two-armed bandit problem where n_0 is large and many observations have been taken on X_1 and relatively few on X_2, and the solution of the one-armed bandit problem for large n_0.

This problem produces the following continuous time analogue.

PROBLEM 18.2. Continuous version of two-armed bandit problem. Let $X_1(t_1)$, $X_2(t_2)$ be Wiener processes with unknown drifts μ_1 and μ_2 per unit time and known constant variance σ^2 per unit time. One is permitted to observe either process, and to switch from one to the other as long as $t_1 + t_2 < n_0$. When the amount n_0 of time is used up the payoff is $X_1(t_1) + X_2(t_2)$. What is the best procedure for switching from one process to the other? Assume that the prior distributions of μ_1 and μ_2 are the independent normals $\mathcal{N}(v_1, \tau_1^2)$ and $\mathcal{N}(v_2, \tau_2^2)$.

Here the Bayes estimates of μ_1 and μ_2 are given by the independent Wiener processes $Y_1(s_1)$ and $Y_2(s_2)$, where

$$(18.6) \qquad s_i^{-1} = (\tau_i^{-2} + t_i\sigma^{-2}), \qquad\qquad i = 1, 2,$$

$$(18.7) \qquad Y_i(s_i) = [v_i\tau_i^{-2} + X_i(t_i)\sigma^{-2}]/(\tau_i^{-2} + t_i\sigma^{-2}), \qquad i = 1, 2,$$

and the payoff when $t_1 + t_2 = n_0$ is a linear function of

$$\frac{Y_1}{s_1} + \frac{Y_2}{s_2} = \left(\frac{Y_1 + Y_2}{2}\right)\left\{\frac{1}{s_1} + \frac{1}{s_2}\right\} + \left(\frac{Y_1 - Y_2}{2}\right)\left\{\frac{1}{s_1} - \frac{1}{s_2}\right\}.$$

No matter what procedure is used the expectation of $(Y_1 + Y_2)(s_1^{-1} + s_2^{-1})$ at the end of sampling is fixed at $(v_1 + v_2)(\tau_1^{-2} + \tau_2^{-2} + n_0\sigma^{-2})$. Thus attention may be confined to maximizing the expectation, at the end of the allotted time, of

$$(18.8) \qquad\qquad Y\{s_1^{-1} - s_2^{-1}\},$$

where $Y = Y_1 - Y_2$ is a process with the following property. If process i is used $E(dY) = 0$, $\mathrm{var}(dY) = -ds_i$. By suitable normalization we may consider the case where the game is stopped at $s_1^{-1} + s_2^{-1} = 1$.

In this version note that if Y is highly positive, one is inclined to reduce s_1, i.e., use X_1, whereas if Y is highly negative one desires to reduce s_2. Arguments similar to ones previously presented suggest that the optimal procedure is defined by a surface $\tilde{y}(s_1, s_2)$ so that X_1 is used when $Y > \tilde{y}(s_1, s_2)$ and X_2 when $Y < \tilde{y}(s_1, s_2)$. The optimal risk $\rho(y, s_1, s_2)$ (negative expected payoff) satisfies the following properties for $s_1^{-1} + s_2^{-1} \leq 1$:

$$\tfrac{1}{2}\rho_{yy} = \rho_{s_1}, \qquad\qquad y \geq \tilde{y},$$

$$\tfrac{1}{2}\rho_{yy} = \rho_{s_2}, \qquad\qquad y \leq \tilde{y},$$

$$\rho(y, s_1, s_2) = (s_2^{-1} - s_1^{-1})y, \qquad s_1^{-1} + s_2^{-1} = 1,$$

$$\rho_y^+ = \rho_y^-, \qquad\qquad y = \tilde{y},$$

$$\rho_{yy}^+ = \rho_{yy}^-, \qquad\qquad y = \tilde{y},$$

where the last equation is an optimality condition.

Formal expansions have been derived for $s_1^{-1} + s_2^{-1} \approx 1$ (near the end of observation) and for s_1 large and s_2 bounded. The latter expansions resemble those for the one-armed bandit problem. To prove that these represent asymptotic approximations to the optimal boundary it is necessary to find means of bounding the solution. One device which would be useful depends on a conjecture.

CONJECTURE. Consider a restricted version of Problem 18.2 where once X_2 is used, one is forced to continue with it. Given the same prior information, if the original (unrestricted) problem calls for starting with X_1, so does the restricted problem.

If this conjecture is true, the solution of the restricted problem provides a bound for Problem 18.2. Moreover, the restricted version is equivalent to a one-armed bandit type of stopping problem. Note that the conjecture is *not* true in discrete time versions of this problem for which the prior distributions make μ_1 and μ_2 dependent.

The two-armed bandit problem has much in common with stochastic control problems and one of the technical difficulties, as compared with stopping problems, is that the risk is not known at the switching boundary.

As far as formulation is concerned, the continuous time version of the two-armed bandit can be generalized easily to m arms. It can also be modified to permit stopping before time n_0 has elapsed if all arms seem unfavorable.

One major point of the bandit problems which has not been stressed in our discussion but which is a key to its interest is that it deals with the question of when one should take an action whose immediate payoff seems to be negative in the hope of getting extra valuable information. Thus in the clinical trials situation we may find ourselves in the ethically difficult situation of using the unknown drug, when current information makes it compare unfavorably with the standard, for fear that it is an improvement which we do not wish to deny prematurely to the *horizon* of $n_0 - N$ remaining patients.

The simple-minded approach of always using the arm which is currently estimated to be better may be easier to evaluate but misses the major issue. Some work on this and alternative schemes such as "pick the winner" (switch treatment when it fails) and finite memory devices appear in [Q1], [Y2], [S5], [C15].

We close this section with a brief description of the backward induction numerical computation applied to the two-armed bandit.

Let $Y(s_1, s_2)$ be defined for positive integers s_1, s_2 so that

$$Y(s_1, s_2 - 1) = Y(s_1, s_2) \pm 1 \quad \text{w.p. } 1/2 \text{ if arm 2 is used,}$$

$$Y(s_1 - 1, s_2) = Y(s_1, s_2) \pm 1 \quad \text{w.p. } 1/2 \text{ if arm 1 is used.}$$

Then

$$\rho(y, s_1, s_2) = \tfrac{1}{2} \min[\rho(y + 1, s_1 - 1, s_2) + \rho(y - 1, s_1 - 1, s_2), \rho(y + 1, s_1, s_2 - 1)$$
$$+ \rho(y - 1, s_1, s_2 - 1)]$$

is the main equation of the backward induction. Special precautions and adjustments were taken for

(i) Scale adjustment: the range used was $s_1^{-1} + s_2^{-1} \geq 2m^{-1}$, m a specified integer. Large values of m yield finer approximations to the continuous problem after normalization. The corresponding normalized values of s_i are $s_i^* = 2s_i/m$.

(ii) End adjustment: the boundary $s_1^{-1} + s_2^{-1} = 2m^{-1}$ does not go through lattice points.

(iii) Boundary effect: to compute $\rho(y, s_1 + r_1, s_2 + r_2)$, one requires the value of $\rho(y + r_1 + r_2, s_1, s_2)$. To avoid calculation for an excessive number of y, one must limit y and calculate the effect of this limitation.

A study of the results of the backward induction for $s_1^* \leq 20$, $s_2^* \leq 20$, $m = 8$ and for $s_1^* \leq 12$, $s_2^* \leq 12$, $m = 32$ led to the following approximation for $\tilde{y} = g(s_1^*, s_2^*)$, the optimal switching boundary in the normalized problem, which leaves much to be desired theoretically but which seems to fit the computational results over the range considered:

(18.9a) $$g(s_1^*, s_2^*) = -g(s_2^*, s_1^*), \qquad\qquad s_1^* \leq s_2^*,$$

(18.9b) $$g(s_1^*, s_2^*) = \sqrt{s_1^*}\, \alpha_1(t'), \qquad\qquad s_2^* \leq 2, s_1^* \geq s_2^*,$$

(18.9c) $$g(s_1^*, s_2^*) = (t')^{-1/2}\alpha_2(t''), \qquad\qquad s_2^* > 2, s_1^* \geq s_2^*,$$

where

(18.9d) $$t' = s_2^*/(s_2^* - 1)s_1^*,$$

(18.9e) $$t'' = s_2^*/s_1^*,$$

and α_1 and α_2 are tabulated in Table 4.

In the nonnormalized version, the optimal procedure is to observe X_1 or X_2 depending on whether

(18.10) $$Y \geq n_1^{-1/2} g(n_1 s_1, n_1 s_2),$$

where Y is the current Bayes estimate of the difference $\mu_1 - \mu_2$ and $n_1 = \tau_1^{-1} + \tau_2^{-1} + n_0\sigma^{-2}$.

19. Sequential estimation of a normal mean. In §3 we derived the optimal sample size for estimating the unknown mean μ of a normal distribution $\mathcal{N}(\mu, \sigma^2)$ with known variance. The optimal sample size for the cost function

$$cn + k(t - \mu)^2$$

was

(19.1) $$n_0 = (k\sigma^2/c)^{1/2},$$

which provided an expected cost of $2(k\sigma^2 c)^{1/2}$ when the sample mean is used as the estimate of μ. We suggested that if σ were not known, one could after each observation, use the current estimate of σ^2 in place of σ^2 in (19.1) to indicate when an adequate sample size had been reached. It was anticipated that using such a simple-minded scheme, the loss due to ignorance would be relatively small and so we turned our attention to the apparently more subtle problems in sequential testing of hypotheses.

Let us return to the sequential estimation problem and determine how much is the cost of ignorance and whether there is much room for improvement in this simple-minded scheme. First let us define the rule more carefully. Since at least two observations are required to estimate σ^2 by conventional means and the typical estimates of variance are highly variable for low sample sizes, we shall modify the procedure to require at least m observations. Our procedure determines the sample size N by

(19.2) $N = \text{least integer} \quad n \geq m \quad \text{for which} \quad n \geq (ks_n^2/c)^{1/2},$

where

$$s_n^2 = (n - 1)^{-1} \sum_{i=1}^{n} (X_i - \overline{X}_n)^2, \quad \overline{X}_n = n^{-1} \sum_{i=1}^{n} X_i$$

and uses $\overline{X}_N$ as the estimate of μ.

The *regret* or *cost of ignorance of* σ^2 is easily seen to be

(19.3) $\omega = cE(N - n_0) + k\sigma^2 E(N^{-1} - n_0^{-1}).$

In connection with a sampling inspection application Robbins [R1] attacked a minor variation of this problem with $(t - \mu)^2$ replaced by $|t - \mu|$ and indicated that ω was remarkably low. His results were generalized to the use of $|t - \mu|^r$ for the cost of error in estimation by Starr and Woodroofe [S1]. They also extended Robbins results to prove the following theorem.

THEOREM 19.1. *For* $m \geq r + 1$, $\omega \leq Kc$, *where* K *is independent of* k, σ *and* c.

In other words the cost of ignorance of σ^2 is bounded by the cost of a finite number of observations irrespective of the other parameters. In conventional statistical applications the cost of ignorance is typically of the order of magnitude of the cost of the square root of the number of observations involved. Here the regret is bounded by the cost of some finite number K of observations no matter how large n_0 is. Simulations carried out by Robbins for $r = 1$, $m = 2$ and 3 indicated that $K = 2$ and $K = 1$ would suffice unless n_0 is very small in which case our $m = 3$ scheme insists on 3 observations, two of which are almost wasted.

Thus there seems to be little room for improvement. Actually, these results are very sensitive to the normality assumption and if that is violated, the method should be adjusted to allow for more *robust* estimates of σ^2.

It should be remarked that a similar theorem was derived previously in the confidence interval context by Simons [S6].

We outline a proof of Theorem 19.1 for the case of squared error loss.

First, by factoring out c and transforming to $X^* = (k/c)^{1/2}X$, we may reduce the problem to a normalized version where $c = 1$, $k = 1$. Then $n_0 = \sigma$ and our procedure calls for stopping for the first $n \geq m$ for which

(19.4) $$\sum_{i=1}^{n} (X_i - \bar{X}_n)^2 \leq (n - 1)n^2.$$

But $\sum_{i=1}^{n} (X_i - \bar{X}_n)^2/\sigma^2$ has a chi-square distribution with $n - 1$ degrees of freedom and (19.4) can be expressed as

(19.4') $\quad S_{n-1} = Y_1 + Y_2 + \cdots + Y_{n-1} \leq (n - 1)n^2/\sigma^2 = (n - 1)n^2/n_0^2,$

where the Y_i are i.i.d. random variables with the distribution of χ^2 with one degree of freedom, i.e., the square of an $\mathcal{N}(0, 1)$ random variable. Note that when n is large $S_{n-1} \approx (n - 1) + O_p(\sqrt{n - 1})$ and S_{n-1} tends to intersect $(n - 1)n^2/n_0^2$ near $n = n_0$. Basically the proof revolves around showing that N cannot be too far away from n_0 with much probability. The case where σ is bounded raises no problem. Let us consider $\sigma = n_0 \to \infty$.

As a preliminary we note that by applying the inequality (9.1.4) to $S_{n-1} - (n - 1)$ $\cdot n^2/n_0^2$ together with the fact that

$$E[e^{tS_{n-1}}] = E[e^{tY_1}]^{n-1} = E[e^{tX_1^2}]^{n-1} = (1 - 2t)^{-(n-1)/2},$$

we obtain for $n \leq n_0$ and all $t < 1/2$,

$$P[S_{n-1} \leq (n - 1)n^2/n_0^2] \leq (1 - 2t)^{-(n-1)/2} \exp[-t(n - 1)n^2/n_0^2].$$

Substituting $1 - 2t = n_0^2/n^2$, we have

(19.5) $$P[S_{n-1} \leq (n - 1)n^2/n_0^2] \leq \left\{ \frac{n}{n_0} \exp\left[\frac{1}{2}\left(1 - \frac{n^2}{n_0^2} \right) \right] \right\}^{n-1}.$$

Thus for $m \geq 3$ and for fixed $\theta, 0 < \theta < 1$,

(19.6) $$P(m \leq N \leq \theta n_0) = O[n_0^{-(m-1)}] \quad \text{as} \quad \sigma \to \infty.$$

Now, in our normalized problem with $\sigma = n_0$,

$$\omega = E[(N - n_0) + n_0^2(N^{-1} - n_0^{-1})] = E\frac{(N - n_0)^2}{N}.$$

For $N \leq \theta n_0$, $(N - n_0)^2 N^{-1} \leq n_0^2/m$ while for $N > \theta n_0$, $(N - n_0)^2 N^{-1} \leq \theta^{-1}$ $\cdot [(N - n_0)^2/n_0]$. Hence,

(19.7) $$\omega \leq \frac{n_0^2}{m} P[N \leq \theta n_0] + \theta^{-1} E\left[\frac{(N - n_0)^2}{n_0} \right].$$

The first term on the right is bounded for $m \geq 3$. Thus it suffices to show that the expectation in the last term is bounded. Let L_1 be the line through $[\theta n_0, (\theta n_0 - 1)\theta^2]$ and $(n_0, n_0 - 1)$ and let L_2 be the line through $(n_0, n_0 - 1)$ tangent to the convex curve $S = (n - 1)n^2/n_0^2$. These have slopes of $(1 + \theta + \theta^2) - (1 + \theta)/n_0$ and $3 - 2/n_0$. Replace these by L_1^* and L_2^*. Let L_1^* be the line through $(n_0, n_0 - 1)$ with slope $1 + \theta + \theta^2$ which intersects the curve at abscissas n_0 and $\theta^* n_0$, $\theta^* = \theta + o(1)$. Let L_2^* be the tangent line to the curve with slope 3 which touches the curve at an abscissa between n_0 and $n_0 + 1$ and lies below the curve elsewhere. Let M_i be the least value of n for which S_{n-1} is below L_i^*, $i = 1, 2$.

If $\theta^* n_0 \leqq N \leqq n_0$, $(N - n_0)^2 \leqq (M_1 - n_0)^2$ and if $N > n_0$, $(N - n_0)^2 \leqq (M_2 - n_0)^2$. Thus

$$E(N - n_0)^2 \leqq E(M_1 - n_0)^2 + E(M_2 - n_0)^2 + n_0^2 P(N \leqq \theta^* n_0).$$

Now S_{n-1} first goes below L_1^* when $\sum_{i=1}^{n-1} [Y_i - (1 + \theta + \theta^2)]$ goes below $-(n_0 - 1)(\theta + \theta^2)$. Each term of this sum has mean $-(\theta + \theta^2) < 0$, and renewal theory [C9] gives the time of first passage as $n_0 + O_p(\sqrt{n_0})$ with variance of order n_0. Thus $E(M_1 - n_0)^2$ is of order n_0. The same reasoning applies to L_2^*. The third term is bounded as before for $m \geqq 3$. Thus ω is bounded.

20. Sequential estimation of the common mean of two normal populations. In this section we consider an application of the last two sections to a simple prototype of a problem in experimental design. We have two instruments available for the measurement of a desired quantity. Both are adjusted to provide unbiased estimates, but the precisions of the two instruments are not known. How should one allocate measurements between the two instruments and when should one stop sampling? In one variation of this problem, one of the two instruments has known precision. In that variation, it makes sense to start with the instrument of unknown precision and to use that with two objectives. One is to estimate the unknown quantity, the other is to measure the precision. If and when it is decided that the unknown instrument is worse than the other, we can switch to the other and finish off the experiment. If this is never decided, ultimately one decides that enough information has been accumulated and terminates sampling.

These problems have been studied by Mallik [M3] and Chernoff [C7] who have observed that the cost due to using the less efficient instrument n times is proportional to n multiplied by the difference in the precisions (when they are close to each other). This observation motivated the application of the solutions of the bandit problems since the cost of ignorance in those problems is also of this form. On the other hand, the cost of taking too few or too many observations is analogous to that of the sequential estimation problem of § 19. Thus the simple-minded stopping rule of that section was also applied in proposing a sequential rule.

Monte Carlo simulations were carried out to estimate the cost of ignorance using the proposed methods in these problems and these were quite moderate. In problems where the known precisions would have called for 32, 100 and 316 observations respectively, the cost of ignorance, measured in observations, peaked at 5, 7 and 10, in the one unknown instrument version; and at 7, 11 and 18, in the two unknown instruments version. These procedures, which are somewhat ad hoc and may bear some improvement, do remarkably well when the instruments have equal precision. There the cost of ignorance was 0.85, 0.92 and 1.21, in the one unknown instrument version and 1.4, 1.8 and 2.3, in the two unknown instruments version. This remarkable reduction suggests that it may be possible to improve performance slightly at the peaks by increasing cost when the precisions are equal.

Both of these methods are based on the assumption of normality. They invoke the behavior of s_n^2, the sample variance as an estimate of variance. The variability of this estimate depends on the fourth moment about the mean of the original random variable. This moment which is $3\sigma^4$ for the normal distribution is notoriously sensitive to the assumption of normality and it is clear that practical application of these methods require modification to use more *robust* estimates of precision.

We proceed to state the two unknown instruments estimation problem formally, and to develop its relation to the bandit and sequential estimation problems which will be used to motivate the procedure some of whose Monte Carlo simulation results were described.

PROBLEM 20.1. Sequential estimation of a mean using two instruments of unknown precision. Let X_i be normally distributed with common unknown mean μ and unknown variances σ_i^2, $i = 1, 2$. If n_i independent observations are taken on X_i, $i = 1, 2$, the mean is estimated by the weighted average

$$(20.1) \qquad \hat{\mu} = (n_1\hat{\sigma}_1^{-2}\overline{X}_{1n_1} + n_2\hat{\sigma}_2^{-2}\overline{X}_{2n_2})/(n_1\hat{\sigma}_1^{-2} + n_2\hat{\sigma}_2^{-2}),$$

where $\overline{X}_{in_i}$ and $\hat{\sigma}_i^2$ are the sample mean and estimated variance respectively of X_i. (We shall use a slight variation of the usual estimate of σ_i^2.) The cost of error in estimation is $k(\hat{\mu} - \mu)^2$ and the cost per observation is c. After each observation one must choose to stop sampling or to take another observation, and in the latter case one must decide between X_1 and X_2. It is desired to select a good sequential sampling rule.

Given a proper Bayesian framework, the problem of finding an optimal procedure is well-defined. We shall merely seek a plausible procedure using Bayesian analysis occasionally as a minor technical aid.

For example if $W_1, W_2, \cdots, W_n$ are independent observations on an $\mathcal{N}(\mu, \sigma^2)$ random variable and $V = \sigma^{-2}$ has the gamma distribution $\Gamma(a, b)$ with density

$$g(v; a, b) = e^{-av}a^b v^{b-1}/\Gamma(b), \qquad v > 0, a > 0, b > 0,$$

and μ is independently normally distributed with infinite variance (ideal flat prior), the posterior density of V can easily be shown to be proportional to

$$g(v; a, b)(v)^{(n-1)/2} \exp\left[-\sum_{i=1}^{n} (W_i - \overline{W}_n)^2 v/2 \right].$$

This implies that the posterior distribution of σ^{-2} is also a gamma distribution. Given the prior $\Gamma(a, b)$ with $a = z_0/2$ and $b = m_0/2$, the posterior is given by

$$\mathcal{L}(\sigma^{-2}|Z_{n-1}) = \Gamma(a', b'),$$

where

$$(20.2) \qquad a' = z/2 = (z_0 + Z_{n-1})/2, \quad b' = m/2 = [m_0 + (n-1)]/2$$

and

$$(20.3) \qquad Z_{n-1} = \sum_{i=1}^{n} (W_i - \overline{W}_n)^2.$$

Thus the prior distribution can conveniently be thought of as being based on $m_0 + 1$ previous observations (m_0 degrees of freedom) with z_0 as the sum of squared deviations from the mean even if m_0 is not an integer.

The mean of the $\Gamma(a, b)$ distribution is easily seen to be b/a. Thus the Bayes estimate of σ^{-2} (for squared error loss) is $\hat{\sigma}^{-2} = m/z$ and

(20.4) $$\hat{\sigma}^2 = z/m = \frac{z_0 + Z_{n-1}}{m_0 + (n-1)}$$

is a reasonable estimate of σ^2. Because the ratio of the σ^2 appear in a natural fashion in our problem, it will be natural to estimate $\log \sigma^2$. For larger values of n, the particular choice of estimate does not matter terribly. But for small values of n, which often play an important role in starting off correctly in our problem, it is natural to use as an estimate of $\log \sigma^2$,

(20.5) $$E(\log \sigma^2 | Z_{n-1}) = \log \hat{\sigma}^2 + \log \frac{m}{2} - \Psi\left(\frac{m}{2}\right),$$

where $\Psi(x) = d[\log \Gamma(x)]/dx$ is the digamma function. This equation and the fact that the conditional variance of $\log \sigma^2$ is $\Psi'(m/2)$ are obtained from the easily computed m.g.f. of $\log \sigma^2$:

$$E(\sigma^{2t} | Z_{n-1}) = (z/2)^t \Gamma\left(\frac{m}{2} - t\right) / \Gamma\left(\frac{m}{2}\right).$$

We return to our problem. Let z_{i_0} and m_{i_0} represent the z_0 and m_0 of the priors of σ_i^{-2}. Let Z_{1,n_1-1} and Z_{2,n_2-1} be the sum of squared deviations from the means for the n_1 and n_2 observations on X_1 and X_2 respectively. Let $\hat{\sigma}_i^2 = z_i/m_i$ with $z_i = z_{i_0} + Z_{i,n_i-1}$ and $m_i = m_{i_0} + (n_i - 1)$ and $\hat{\mu}$ as in (20.1). By applying a scale transformation to X_i we may assume $k = 1$ with no loss of generality.

First we consider a stopping rule in the spirit of § 19. If σ_1 and σ_2 were known, an appropriate estimate of μ would be $(n_1 \sigma_1^{-2} \bar{X}_{1n_1} + n_2 \sigma_2^{-2} \bar{X}_{2n_2})/(n_1 \sigma_1^{-2} + n_2 \sigma_2^{-2})$ which would yield a risk of

$$c(n_1 + n_2) + (n_1 \sigma_1^{-2} + n_2 \sigma_2^{-2})^{-1}.$$

If $\sigma_1 < \sigma_2$ we may ask how large should n_1 be made (keeping n_2 fixed) to minimize the risk. Minimizing with respect to n_1 gives us

$$n_1 \sigma_1^{-2} + n_2 \sigma_2^{-2} = c^{-1/2} \sigma_1^{-1},$$

where the left-hand side of the above equation represents the amount of information cumulated for estimating μ. There is no point to taking more observations if the left-hand side exceeds the right-hand side. Thus we suggest that *sampling be stopped as soon as*

(20.6) $$n_1 \hat{\sigma}_1^{-2} + n_2 \hat{\sigma}_2^{-2} \geqq c^{-1/2} \max(\hat{\sigma}_1^{-1}, \hat{\sigma}_2^{-1}),$$

as long as $n_1 + n_2$ exceeds some minimum sample size n_0.

Having stopped sampling after n_i observations on X_i, $i = 1, 2$, we use $\hat{\mu}$ as in (20.1). Given $\hat{\sigma}_1^2$ and $\hat{\sigma}_2^2$, the conditional risk (since $\overline{X}_{in_i}$ is independent of $\hat{\sigma}_i$) is

$$R = c(n_1 + n_2) + \frac{n_1\hat{\sigma}_1^{-4}\sigma_1^2 + n_2\hat{\sigma}_2^{-4}\sigma_2^2}{(n_1\hat{\sigma}_1^{-2} + n_2\hat{\sigma}_2^{-2})^2},$$

which we shall compare with the target $2c^{1/2}\sigma_1$. The target is based on the assumption that $\sigma_1 < \sigma_2$ are known and the optimal numbers $n_{10} = c^{-1/2}\sigma_1$, $n_{20} = 0$ are used with resulting risk $cn_{10} + \sigma_1^2/n_{10} = 2c^{1/2}\sigma_1$. Assuming that stopping takes place when equality is satisfied in (20.6) with $\hat{\sigma}_1 \leqq \hat{\sigma}_2$, and letting

(20.7) $\hat{N} = \hat{\sigma}_1 c^{-1/2}$,

we have

$$R = \frac{[\hat{N} + (n_1 - \hat{N})]\hat{\sigma}_1^{-2}\left(\dfrac{\sigma_1^2}{\hat{\sigma}_1^2} - 1\right) + n_2\hat{\sigma}_2^{-2}\left(\dfrac{\sigma_2^2}{\hat{\sigma}_2^2} - 1\right)}{(n_1\hat{\sigma}_1^{-2} + n_2\hat{\sigma}_2^{-2})^2} + (n_1\hat{\sigma}_1^{-2} + n_2\hat{\sigma}_2^{-2})^{-1}$$

$$+ c[\hat{N} + (n_1 - \hat{N}) + n_2],$$

$$R = c\hat{\sigma}_1^2\left[(n_1 - \hat{N})\hat{\sigma}_1^{-2}\left(\frac{\sigma_1^2}{\hat{\sigma}_1^2} - 1\right) + n_2\hat{\sigma}_2^{-2}\left(\frac{\sigma_2^2}{\hat{\sigma}_2^2} - 1\right)\right] + c[(n_1 - \hat{N}) + n_2]$$

$$+ c^{1/2}\hat{\sigma}_1\left[\frac{\sigma_1^2}{\hat{\sigma}_1^2} - 1 + 2\right].$$

Replacing $\hat{\sigma}_i^2$ by

$$\sigma_i^2[1 + (\hat{\sigma}_i^2 - \sigma_i^2)/\sigma_i^2] = \sigma_i^2[1 + \eta_i],$$

the coefficient of $c^{1/2}$ in R is

$$\hat{\sigma}_1^{-1}[\sigma_1^2 + \hat{\sigma}_1^2] = \sigma_1^{-1}(1 + \eta_1)^{-1/2}\sigma_1^2(2 + \eta_1)$$

$$= \sigma_1[2 + \tfrac{1}{4}\eta_1^2 + \cdots].$$

Since the stopping rule equality implies $(\hat{N} - n_1)\hat{\sigma}_1^{-2} = n_2\hat{\sigma}_2^{-2}$, the coefficient of c in R is

$$\hat{\sigma}_1^2\left[(n_1 - \hat{N})\hat{\sigma}_1^{-2}\left(\frac{\sigma_1^2}{\hat{\sigma}_1^2} - 1\right) + n_2\hat{\sigma}_2^{-2}\left(\frac{\sigma_2^2}{\hat{\sigma}_2^2} - 1\right)\right] + [(n_1 - \hat{N}) + n_2]$$

$$= n_2\left[\frac{-\hat{\sigma}_1^2}{\hat{\sigma}_2^2}\left(\frac{\sigma_1^2}{\hat{\sigma}_1^2} - 1\right) + \frac{\hat{\sigma}_1^2}{\hat{\sigma}_2^2}\left(\frac{\sigma_2^2}{\hat{\sigma}_2^2} - 1\right) - \frac{\hat{\sigma}_1^2}{\hat{\sigma}_2^2} + 1\right]$$

$$= n_2\left[1 - \frac{\sigma_1^2}{\sigma_2^2} - \frac{\sigma_1^2}{\sigma_2^2}\eta_2(\eta_1 - \eta_2) + \cdots\right].$$

Let $\omega = R - 2c^{1/2}\sigma_1$ be the *conditional regret* or conditional cost of ignorance of σ_1 and σ_2. Then

$$(20.8) \quad \omega = cn_2\left[1 - \frac{\sigma_1^2}{\sigma_2^2}\right] + c^{1/2}[\tfrac{1}{4}\sigma_1\eta_1^2 + \cdots] - cn_2\frac{\sigma_1^2}{\sigma_2^2}[\eta_2(\eta_1 - \eta_2) + \cdots].$$

When the number of observations on X_1 is of the order of $c^{-1/2}$ we would ordinarily expect η_1 to be of the order of magnitude $c^{1/4}$ and the second term in (20.8) would be of the order of c and the main term in ω is essentially given by

$$(20.9) \quad cn_2[1 - \sigma_1^2/\sigma_2^2] \quad \text{when} \quad \sigma_1^2 < \sigma_2^2.$$

The case when $\sigma_1^2 < \sigma_2^2$ but $\hat{\sigma}_1^2 > \hat{\sigma}_2^2$ involves more careful analysis. This possibility is of concern when σ_1^2 and σ_2^2 are relatively close together (i.e., within $O(c^{1/4})$) and then one expects both n_1 and n_2 to be $O_p(c^{-1/2})$. One may still derive the same major contribution to risk. (The interested reader may find it more convenient to apply symmetry after treating the case $\sigma_1^2 > \sigma_2^2$ and $\hat{\sigma}_1^2 < \hat{\sigma}_2^2$ for which most of the above algebra remains unchanged though ω must be replaced by $R - 2c^{1/2}\sigma_2$.)

Since the case where σ_1^2 and σ_2^2 are relatively close is the one requiring sophisticated treatment and σ_2^2/σ_1^2 appears as a basic parameter in this problem, let

$$(20.10) \quad v_i = -\log \sigma_i^2.$$

Then $1 - \sigma_1^2/\sigma_2^2 \approx v_1 - v_2$ and the conditional loss due to ignorance is asymptotically

$$(20.9') \quad cn_2(v_1 - v_2).$$

If we think of the observations as providing information on v_1 and v_2, the *loss due to ignorance relates to that of the two-armed bandit problem* (see equation (18.5)) except that the data are not quite independent *normally* distributed observations with mean v_i and constant known variance.

Insofar as the estimate of v_i based on successive observations on X_i resembles a process of successive means of i.i.d. normal random variables, it seems reasonable to apply the solution of the two-armed bandit problem here.

Applying (20.5) we use as our estimate of $-v_i$,

$$(20.11) \quad Y_i = \log \hat{\sigma}_i^2 + \log \frac{m_i}{2} - \Psi\left(\frac{m_i}{2}\right)$$

and the role of s_i, the reciprocal of the cumulated information or precision, in the bandit problem, is played by the variance of the conditional estimate of v_i, i.e.,

$$(20.12) \quad s_i = \Psi'(m_i/2) \approx 2/m_i.$$

The potential total amount of information anticipated, n_0^*, can be estimated. If $\hat{\sigma}_1^2 < \hat{\sigma}_2^2$ where these are based on n_1 and n_2 observations respectively, we anticipate taking extra observations on X_1, i.e., increasing n_1 till

$$c(n_1 + n_2) + (n_1\hat{\sigma}_1^{-2} + n_2\hat{\sigma}_2^{-2})^{-1}$$

is minimized. This occurs when $n_1\hat\sigma_1^{-2} + n_2\hat\sigma_2^{-2} = c^{-1/2}\hat\sigma_1^{-1}$. Thus we anticipate the eventual value of n_1 to be

$$(20.13) \qquad n_1^* = (c^{-1/2}\hat\sigma_1^{-1} - n_2\hat\sigma_2^{-2})\hat\sigma_1^2.$$

The corresponding values of the s_i are $\Psi'[(n_1^* - 1 + m_{10})/2]$ and $\Psi'[(n_2 - 1 + m_{20})/2]$ whose inverses approximately sum to

$$(20.14) \qquad n_0^* = \tfrac{1}{2}[m_{10} + m_{20} - 2 + c^{-1/2}\hat\sigma_1 + n_2(1 - \hat\sigma_1^2/\hat\sigma_2^2)].$$

Thus when sampling is continued we select X_1 or X_2 according to whether or not

$$(20.15) \qquad Y = Y_2 - Y_1 \gtreqless (n_0^*)^{-1/2}g(n_0^*s_1, n_0^*s_2),$$

where g is the function representing our approximation to the solution of the normalized version of the two-armed bandit problem.

One minor complication exists. The stopping rule (20.6) does not imply $s_1^{-1} + s_2^{-1} \leq n_0^*$ and g is not defined otherwise. Thus we may impose a supplementary stopping rule which is asymptotically equivalent to (20.6): *stop if*

$$(20.16) \qquad s_1^{-1} + s_2^{-1} \geq n_0^*.$$

We recapitulate our procedure. Having normalized the problem so that $k = 1$, we select m_{i0}, z_{i0} and the minimum sample size n_0 distributed evenly between X_1 and X_2. Having taken n_i observations on $X_i, i = 1, 2$, we compute $m_i, z_i, \hat\sigma_i^2, Y_i$ and s_i. Suppose $\hat\sigma_1 < \hat\sigma_2$. Then we compute n_0^*.

Sampling is stopped if

$$(20.6) \qquad n_1\hat\sigma_1^{-2} + n_2\hat\sigma_2^{-2} \geq c^{-1/2}\hat\sigma_1^{-1}$$

or if

$$(20.16) \qquad s_1^{-1} + s_2^{-1} \geq n_0^*$$

unless $n_1 + n_2 < n_0$. If sampling is continued, select X_1 or X_2 according to whether or not

$$(20.15) \qquad Y_2 - Y_1 \gtreqless (n_0^*)^{-1/2}g(n_0^*s_1, n_0^*s_2).$$

A similar development, differing somewhat in details, was presented by Mallik [M3] in his dissertation devoted to the one-armed bandit version of this problem where one of the instruments has known precision.

In this and a number of other problems, we have applied normal theory results to nonnormal problems. Under more general circumstances, the derivative of the logarithm of the likelihood evaluated at the boundary point (separating the hypotheses) closest to θ_n behaves asymptotically like a Gaussian process of independent increments which carries the essential information on the unknown parameter. It seems plausible that our methods will lead to effective results if we replace $\sum X_i$ in ordinary problems by this derivative. To prove such theorems would require the use of a theorem on convergence of measures of stochastic processes. One step in such a direction has been taken by Breslow [B9].

21. Conclusion. These lectures were given structure by the emphasis on those ideas in sequential analysis and optimal design of experiments to which I have devoted a substantial amount of effort over the years. In pursuing these ideas, digressions into major principles of testing hypotheses and estimation were required. On the other hand, many fundamental and important results in directions related to the main theme were neglected.

We end with a list of a few of these which have been widely quoted and extensively developed. The concepts of Pitman efficiency [P1] and Bahadur efficiency [B12] in testing hypotheses are meaningful and useful in the efficient design of experiments for hypotheses with nearby alternatives. Stochastic approximation originating with the Robbins–Monro method [R5] and the response-surface methods of Box–Wilson [B13] have many elements in common with one another and with sequential design.

Anscombe's work in sequential estimation and clinical trials [A5]–[A7] anticipated some of the heuristics of asymptotic theory applied to sequential analysis and to the one-armed bandit problem, to which he also contributed.

REFERENCES

[A1] A. ALBERT, *The Penrose Moore Pseudo Inverse with Statistical Applications*, Academic Press, New York, to appear.

[A2] A. E. ALBERT, *The sequential design of experiments for infinitely many states of nature*, Ann. Math. Statist., 32 (1961), pp. 774–799.

[A3] K. J. ARROW, D. BLACKWELL AND M. A. GIRSHICK, *Bayes and minimax solutions of sequential design problems*, Econometrica, 17 (1949), pp. 213–244.

[A4] G. R. ANTELMAN, *Insensitivity to non-optimal design in Bayesian decision theory*, J. Amer. Statist. Assoc., 60 (1965), pp. 584–601.

[A5] F. J. ANSCOMBE, *Notes on sequential sampling plans*, J. Royal Statist. Soc. Ser. A, 123 (1960), pp. 297–306.

[A6] ———, *Sequential medical trials*, J. Amer. Statistic. Assoc., 58 (1963), pp. 365–387.

[A7] ———, *Large-sample theory of sequential estimation*, Proc. Cambridge Philos. Soc., 48 (1952), pp. 600–607.

[A8] T. W. ANDERSON AND R. R. BAHADAR, *Classification into two multivariate normal distributions with different covariance matrices*, Ann. Math. Statist., 33 (1962), pp. 420–431.

[B1] D. BLACKWELL, *The range of certain vector integrals*, Proc. Amer. Math. Soc., 2 (1951), pp. 390–395.

[B2] S. BESSLER, *Theory and applications of the sequential design of experiments, k-actions and infinitely many experiments: Part I—Theory*, Tech. Rep. 55, Department of Statistics, Stanford University, California, 1960.

[B3] ———, *Theory and applications of the sequential design of experiments, k-actions and infinitely many experiments: Part II—Applications*, Tech. Rep. 56, Department of Statistics, Stanford University, California, 1960.

[B4] G. E. P. BOX AND W. J. HILL, *Discrimination among mechanistic models*, Technometrics, 9 (1967), pp. 57–71.

[B5] W. J. BLOT AND D. A. MEETER, *Sequential experimental design procedure*, J. Amer. Statist. Assoc., 68 (1973), pp. 586–593.

[B6] J. A. BATHER, *Bayes procedure for deciding the sign of a normal mean*, Proc. Cambridge Philos. Soc., 58 (1962), pp. 599–620.

[B7] T. L. BOULLION AND P. L. ODELL, *Generalized Inverse Matrices*, John Wiley, New York, 1971.

[B8] J. V. BREAKWELL AND H. CHERNOFF, *Sequential tests for the mean of a normal distribution II (large t)*, Ann. Math. Statist., 35 (1964), pp. 162–173.

[B9] N. BRESLOW, *On large sample sequential analysis with applications to survivorship data*, J. Appl. Probability, 6 (1969), pp. 261–274.

[B10] J. BATHER, *Optimal stopping problems for Brownian motion*, Advance Appl. Probability, 2 (1970), pp. 259–286.

[B11] S. BESSLER, H. CHERNOFF AND A. W. MARSHALL, *An optimal sequential accelerated life test*, Technometrics, 4 (1962), pp. 367–379.

[B12] R. R. BAHADUR, *Stochastic comparison of tests*, Ann. Math. Statist., 31 (1960), pp. 276–295.

[B13] G. E. P. BOX AND K. P. WILSON, *On the experimental attainment of optimum conditions*, J. Royal Statist. Soc. Ser. B, 13 (1951), pp. 1–45.

[C1] H. CHERNOFF, *Large sample theory, parametric case*, Ann. Math. Statist., 27 (1956), pp. 1–22.

[C2] W. COMMINS, JR., *Asymptotic variance as an approximation to expected loss for maximum likelihood estimates*, Tech. Rep. 46, Department of Statistics, Stanford University, California 1959.

[C3] D. G. CHAPMAN AND R. ROBBINS, *Minimum variance estimation without regularity assumptions*, Ann. Math. Statist., 22 (1951), pp. 581–586.

[C4] H. CHERNOFF, *Locally optimal designs for estimating parameters*, Ibid., 24 (1953), pp. 586–602.

[C5] ———, *A measure of asymptotic efficiency for tests of a hypothesis based on the sum of observations*, Ibid., 23 (1952), pp. 493–507.

[C6] H. CRAMÉR, *Sur un nouveau théorème-limite de la théorie des probabilités*, Actualités Sci. Ind., F36, Paris, 1938.

[C7] H. CHERNOFF, *The efficient estimation of a parameter measurable by two instruments of unknown precisions*, Optimizing methods in Statistics, J. S. Rustagi, ed., Academic Press, New York, 1971.

[C8] ———, *Sequential design of experiments*, Ann. Math. Statist., 30 (1959), pp. 755–770.

[C9] Y. S. CHOW, H. ROBBINS AND D. SIEGMUND, *Great Expectations: The Theory of Optimal Stopping*, Houghton Mifflin, New York, 1971.

[C10] H. CHERNOFF, *Optimal stochastic control*, Sankhyā Ser. A, 30 (1968), pp. 221–252.

[C11] ———, *Sequential tests for the mean of a normal distribution*, Proc. Fourth Berkeley Symp. Math. Statist., 1 (1961), pp. 79–91.

[C12] ———, *Sequential tests for the mean of a normal distribution III (small t)*, Ann. Math. Statist., 36 (1965), pp. 28–54.

[C13] ———, *Sequential tests for the mean of a normal distribution IV (discrete case)*, Ibid., 36 (1965), pp. 55–68.

[C14] H. CHERNOFF AND S. N. RAY, *A Bayes sequential sampling inspection plan*, Ibid., 36 (1965), pp. 1387–1407.

[C15] T. M. COVER AND M. E. HELLMAN, *Learning with finite memory*, Ibid., 41 (1970), pp. 765–782.

[C16] H. CHERNOFF, *Sequential models for clinical trials*, Proc. Fifth Berkeley Symp. Math. Statist., 4 (1967), pp. 805–812.

[C17] ———, *Sequential designs*, The Design of Computer Simulation Experiments, T. H. Naylor, ed., Duke University Press, Durham, NC, 1969.

[C18] ———, *Some measures for discriminating between normal multivariate distributions with unequal covariance matrices*, Multivariate Analysis III, Krishnaiah, ed., Academic Press, New York, 1993, pp. 337–344.

[C19] C. W. CLUNIES-ROSS AND R. H. RIFFENBURGH, *Geometry and linear discrimination*, Biometrika, 47 (1960), pp. 185–189.

[C20] H. CHERNOFF AND A. J. PETKAU, *An optimal stopping problem for sums of dichotomous random variables*, Ann. Probab., 1 (1976), pp. 875–889.

[D1] J. L. DOOB, *Stochastic Processes*, John Wiley, New York, 1953.

[D2] A. DVORETZKY, J. KIEFER AND J. WOLFOWITZ, *Sequential decision problems for processes with continuous time parameter. Testing hypotheses*, Ann. Math. Statist., 24 (1953), pp. 254–264.

[D3] E. B. DYNKIN, *The optimum choice of the instant for stopping a Markov process*, Dokl. Akad. Nauk SSSR, 150 (1963), pp. 238–240.

[E1] G. ELFVING, *Optimum allocation in linear regression theory*, Ann. Math. Statist., 23 (1952), pp. 255–262.

[F1] W. FELLER, *An Introduction to Probability Theory and Its Applications*, vol. 2, John Wiley, New York, 1966.

[G1] P. G. GUEST, *The spacing of observations in polynomial regression*, Ann. Math. Statist., 29 (1958), pp. 294–299.

[H1] H. HOTELLING, *Some improvements in weighing and other experimental techniques*, Ibid., 15 (1944), pp. 297–306.

[J1] V. M. JOSHI, *Asymptotic risk of maximum likelihood estimates*, J. Indian Statist. Assoc., 6 (1968), pp. 31–42.

[K1] J. KIEFER AND J. WOLFOWITZ, *Optimum designs in regression problems*, Ann. Math. Statist., 30 (1959), pp. 271–294.

[K2] S. KARLIN AND W. J. STUDDEN, *Tchebycheff Systems with Applications in Analysis and Statistics*, Interscience, New York, 1966.

[K3] J. KIEFER AND J. SACKS, *Asymptotically optimal sequential inference and design*, Ann. Math. Statist., 34 (1963), pp. 705–750.

[K4] J. KIEFER, *Optimum designs in regression problems, II*, Ibid., 32 (1961), pp. 298–325.

[K5] J. KIEFER AND L. WEISS, *Some properties of generalized sequential probability ratio tests*, Ibid., 28 (1957), pp. 57–74.

[L1] D. R. LUCE AND H. RAIFFA, *Games and Decisions*, John Wiley, New York, 1957.

[L2] L. LECAM, *On some asymptotic properties of maximum likelihood estimates and related Bayes estimates*, Univ. Calif. Publ. Statist., 1 (1953), pp. 277–330.

[M1] T. K. MATTHES, *On the optimality of sequential probability ratio tests*, Ann. Math. Statist., 34 (1963), pp. 18–21.

[M2] D. MEETER, W. PIRIE AND W. BLOT, *A comparison of two model-discrimination criteria*, Technometrics, 12 (1970), pp. 457–470.

[M3] A. K. MALLIK, *Sequential estimation of the common mean of two normal populations*, Tech. Rep. 42, Department of Statistics, Stanford University, California, 1971.

[P1] E. J. G. PITMAN, *Non-parametric statistical inference*, (mimeographed lecture notes), University of North Carolina Inst. of Statist., 1948.

[Q1] K. QUISEL, *Extensions of the two-armed bandit and related processes with on-line experimentation*, Tech. Rep. 137, Department of Statistics, Stanford University, California, 1965.

[R1] H. ROBBINS, *Sequential estimation of the mean of a normal population*, Probability and Statistics, U. Grenander, ed., John Wiley, New York, 1959, pp. 235–245.

[R2] C. R. RAO AND S. K. MITRA, *Generalized Inverse of Matrices and Its Applications*, John Wiley, New York, 1971.

[R3] H. RUBIN AND J. SETHURAMAN, *Bayes risk efficiency*, Sankhyā Ser. A, 27 (1965), pp. 347–356.

[R4] H. RAIFFA AND R. SCHLAIFER, *Applied Statistical Decision Theory*, Division of Research, Graduate School of Business Administration, Harvard University, Massachusetts, 1961.

[R5] H. ROBBINS AND S. MONRO, *A stochastic approximation method*, Ann. Math. Statist., 22 (1951), pp. 400–407.

[S1] N. STARR AND M. B. WOODROOFE, *Remarks on sequential point estimation*, Proc. Nat. Acad. Sci., 63 (1969), pp. 285–288.

[S2] H. SCHEFFE, *The Analysis of Variance*, John Wiley, New York, 1959.

[S3] W. J. STUDDEN, *Elfving's theorem and optimal design for quadratic loss*, Ann. Math. Statist., 42 (1971), pp. 1613–1621.

[S4] A. N. SHIRYAEV, *Statistical Sequential Analysis*, Moscow, 1969. (In Russian.)

[S5] M. SOBEL AND G. H. WEISS, *Play-the-winner sampling for selecting the better of two binomial populations*, Biometrika, 57 (1970), pp. 357–365.

[S6] G. SIMONS, *On the cost of not knowing the variance when making a fixed-width confidence interval for the mean*, Ann. Math. Statist., 39 (1968), pp. 1946–1952.

[S7] G. SCHWARZ, *Asymptotic shapes of Bayes sequential testing regions*, Ibid., (1962), pp. 224–236.

[T1] M. E. THOMPSON, *Continuous parameter optimal stepping problems*, Z. Wahrscheinlichkeitstheorie und Verw. Gebiete, 19 (1971), pp. 302–318.

[W1] A. WALD, *Note on the consistency of the maximum likelihood estimate*, Ann. Math. Statist., 20 (1949), pp. 595–601.

[W2] ———, *Sequential Analysis*, John Wiley, New York, 1947.

[W3] A. WALD AND J. WOLFOWITZ, *Optimum character of the sequential probability ratio test*, Ann. Math. Statist., 19 (1948), pp. 326–339.

[W4] P. WHITTLE, *Some general results in sequential design*, Biometrika, 51 (1964), pp. 123–141.

[Y1] F. YATES, *Complex experiments*, J. Royal Statist. Soc., 2 (1935), pp. 181–247, including discussion. The reference is to p. 211.

[Y2] S. J. YAKOWITZ, *Mathematics of Adaptive Control Processes*, American Elsevier, New York, 1969.

[Z1] S. ZACKS, *The Theory of Statistical Inference*, John Wiley, New York, 1971.